Y0-EPI-402
NORTH DAKOTA
STATE UNIVERSITY
SEP 17 1984
SERIALS DEPT.
LIBRARY

WITHDRAWN

Memoirs of the American Mathematical Society
Number 309

André Joyal
and Myles Tierney

An extension of the Galois theory of Grothendieck

Published by the
AMERICAN MATHEMATICAL SOCIETY
Providence, Rhode Island, USA

September 1984 · Volume 51 · Number 309 (third of 4 numbers)

MEMOIRS of the American Mathematical Society

This journal is designed particularly for long research papers (and groups of cognate papers) in pure and applied mathematics. It includes, in general, longer papers than those in the TRANSACTIONS.

Mathematical papers intended for publication in the Memoirs should be addressed to one of the editors. Subjects, and the editors associated with them, follow:

Ordinary differential equations, partial differential equations and applied mathematics to JOEL A. SMOLLER, Department of Mathematics, University of Michigan, Ann Arbor, MI 48109.

Complex and harmonic analysis to LINDA PREISS ROTHSCHILD, Department of Mathematics, University of California at San Diego, LaJolla, CA 92093

Abstract analysis to WILLIAM B. JOHNSON, Department of Mathematics, Ohio State University, Columbus, OH 43210

Algebra, algebraic geometry and number theory to LANCE W. SMALL, Department of Mathematics, University of California at San Diego, LaJolla, CA 92093

Logic, set theory and general topology to KENNETH KUNEN, Department of Mathematics, University of Wisconsin, Madison, WI 53706

Topology to WALTER D. NEUMANN, Mathematical Sciences Research Institute, 2223 Fulton Street, Berkeley, CA 94720

Global analysis and differential geometry to TILLA KLOTZ MILNOR, Department of Mathematics, Hill Center, Rutgers University, New Brunswick, NJ 08903

Probability and statistics to DONALD L. BURKHOLDER, Department of Mathematics, University of Illinois, Urbana, IL 61801

Classical analysis to PETER W. JONES, Department of Mathematics, University of Chicago, Chicago, IL 60637

Combinatorics and number theory to RONALD GRAHAM, Mathematical Sciences Research Center, AT&T Bell Laboratories, 600 Mountain Avenue, Murray Hill, NJ 07974

All other communications to the editors should be addressed to the Managing Editor, R. O. WELLS, JR., Department of Mathematics, Rice University, Houston, TX 77251

MEMOIRS are printed by photo-offset from camera-ready copy fully prepared by the authors. Prospective authors are encouraged to request booklet giving detailed instructions regarding reproduction copy. Write to Editorial Office, American Mathematical Society, P.O. Box 6248, Providence, Rhode Island 02940. For general instructions, see last page of Memoir.

SUBSCRIPTION INFORMATION. The 1984 subscription begins with Number 289 and consists of six mailings, each containing one or more numbers. Subscription prices for 1984 are $148 list; $74 member. A late charge of 10% of the subscription price will be imposed upon orders received from nonmembers after January 1 of the subscription year. Subscribers outside the United States and India must pay a postage surcharge of $10; subscribers in India must pay a postage surcharge of $15. Each number may be ordered separately; *please specify number* when ordering an individual number. For prices and titles of recently released numbers, refer to the New Publications sections of the NOTICES of the American Mathematical Society.

BACK NUMBER INFORMATION. For back issues see the AMS Catalogue of Publications.

TRANSACTIONS of the American Mathematical Society

This journal consists of shorter tracts which are of the same general character as the papers published in the MEMOIRS. The editorial committee is identical with that for the MEMOIRS so that papers intended for publication in this series should be addressed to one of the editors listed above.

Subscriptions and orders for publications of the American Mathematical Society should be addressed to American Mathematical Society, P. O. Box 1571, Annex Station, Providence, R. I. 02901. *All orders must be accompanied by payment.* Other correspondence should be addressed to P. O. Box 6248, Providence, R. I. 02940.

MEMOIRS of the American Mathematical Society (ISSN 0065-9266) is published bimonthly (each volume consisting usually of more than one number) by the American Mathematical Society at 201 Charles Street, Providence, Rhode Island 02904. Second Class postage paid at Providence, Rhode Island 02940. Postmaster: Send address changes to Memoirs of the American Mathematical Society, American Mathematical Society, P. O. Box 6248, Providence, RI 02940.

Copyright © 1984, American Mathematical Society

The paper used in this journal is acid-free and falls within the guidelines established to ensure permanence and durability.

QA
3
A57
no.309

TABLE OF CONTENTS

Introduction iv

Chapter I - Sup-lattices 1
1. Definitions and duality 1
2. Limits and colimits 2
3. Free sup-lattices 3
4. Sub and quotient lattices 4
5. Tensor products 5

Chapter II - Rings, modules and descent 7
1. Rings and modules 7
2. Tensor product of modules 8
3. Change of rings 9
4. Flatness, projectivity, and purity 11
5. Descent theory for modules 12

Chapter III - Locales 21
1. Locales and commutative monoids 21
2. Limits and colimits 21
3. The free locale 22
4. Local operators and quotients 22
5. The splitting locale 25

Chapter IV - Spaces 27
1. Subspaces 27
2. Points and discrete spaces 28
3. The Sierpinski space 29
4. Pullbacks and projective limits 30
5. The splitting space 31

Chapter V - Open maps of spaces 33
1. Open maps - definition 33
2. Open subspaces 34
3. Conditions for openness 35
4. Open surjections, pullbacks 38
5. A characterization of discrete spaces 40

Chapter VI - Change of base 42
1. Change of base for sup-lattices and locales 42
2. Determination of $s\ell(S^{\underline{A}^{op}})$ and $loc(S^{\underline{A}^{op}})$ 46
3. Determination of $s\ell(sh(A))$ and $loc(sh(A))$ for $A \in loc(S)$ . . . 49
4. The Beck-Chevalley conditions 52
5. The spatial reflection 53

Chapter VII - Open morphisms of topoi 56
1. Open geometric morphisms 56
2. A site characterization of openness 58
3. The spatial cover 59
4. A characterization of atomic topoi 61

Chapter VIII - Descent morphisms of topoi 64
1. Effective descent morphisms 64
2. Open surjections 65
3. Applications to the structure of topoi 68

References 71

ABSTRACT

In this paper we compare, in a precise way, the concept of Grothendieck topos to the classical notion of topological space. The comparison takes the form of a two-fold extension of the idea of space.

Firstly, in classical topology, a space is a set X equipped with a topology of open sets $\mathcal{O}(X) \subseteq P(X)$. Here, we replace $\mathcal{O}(X)$ by an arbitrary complete lattice satisfying the distributive law $u \wedge (\bigvee_{i \in I} u_i) = \bigvee_{i \in I}(u \wedge u_i)$. Such a lattice is called a locale. The concept of sheaf on a locale is clear, and gives rise to a corresponding topos. The category of (extended) spaces and continuous maps is the dual of the category of locales. We study this category systematically, developing particularly the concept of open mapping.

Secondly, we show that the difference between an arbitrary Grothendieck topos and our new notion of space lies in the possibility of action by a spatial groupoid. That is, if $G_1 \rightrightarrows G_0$ is a groupoid in the category of (extended) spaces, then the general notion of Grothendieck topos is captured by considering sheaves on G_0 with a continuous action by G_1. This is an extension of Grothendieck's interpretation of classical Galois theory.

The basic technique is descent theory for morphisms of locales, developed in the general set theory of an arbitrary elementary topos.

AMS(MOS) subject classification (1980): 13B05, 13C99, 18B25, 18B30, 18F10

Library of Congress Cataloging in Publication Data

Joyal, André.
An extension of the Galois theory of Grothendieck.

(Memoirs of the American Mathematical Society ; no. 309)
Bibliography: p.
1. Toposes. 2. Galois theory. 3. Theory of descent (Mathematics) I. Tierney, Myles. II. Title.
III. Series.
QA3.A57 no. 309 [QA169] 510s [512'.32] 84-14587
ISBN 0-8218-2312-4

Copying and reprinting. Individual readers of this publication, and nonprofit libraries acting for them, are permitted to make fair use of the material, such as to copy an article for use in teaching or research. Permission is granted to quote brief passages from this publication in reviews provided the customary acknowledgement of the source is given.

Republication, systematic copying, or multiple reproduction of any material in this publication (including abstracts) is permitted only under license from the American Mathematical Society. Requests for such permission should be addressed to the Executive Director, American Mathematical Society, P.O. Box 6248, Providence, Rhode Island 02940.

The owner consents to copying beyond that permitted by Sections 107 or 108 of the U.S. Copyright Law, provided that a fee of $1.00 plus $.25 per page for each copy be paid directly to Copyright Clearance Center, Inc., 21 Congress Street, Salem, Massachusetts 01970. When paying this fee please use the code 0065-9266/84 to refer to this publication. This consent does not extend to other kinds of copying, such as copying for general distribution, for advertising or promotion purposes, for creating new collective works or resale.

INTRODUCTION

Attempting to define a "Weil cohomology" with the formal properties necessary to establish the Weil conjectures, Grothendieck discovered étale cohomology, a fusion of ordinary sheaf cohomology and Galois cohomology. The definition required an extension of the concept of sheaf to the idea of a sheaf on a site - a category equipped with an a priori notion of covering. Contrary to the topological case, two non-isomorphic sites can give rise to the same category of sheaves. One is thus led to consider the category of sheaves as the natural object of study, rather than the generating site. In this way the notion of a Grothendieck topos - the concept abstracted from categories of sheaves on sites - was introduced. A systematic study of sites, sheaves, and topoi was carried out by M. Artin, A. Grothendieck, and J.L. Verdier in SGA4 (1963/64) [2]. The theory that arose from this study has proved to be extremely useful in algebraic geometry, contributing in an essential way to the complete proof by Deligne of all the Weil conjectures. Of course as Deligne in SGA4½ points out, only a fraction of the general theory is strictly necessary to establish these results - a remark true of any application of a general theory to a special case.

The development of the theory of topos was guided by the analogy between the category of sheaves on a site, and sheaves on an ordinary topological space. Thus a topos was seen to be a kind of generalized space. To quote from SGA4 [2], p. 301:

> "Comme le terme de topos lui-même est censé precisément le suggérer, il semble raisonnable et légitime aux auteurs du présent séminaire de considérer que l'objet de la Topologie est l'étude des topos (et non des seuls espaces topologiques)."

In this paper we intend to deepen this analogy by establishing a theorem, which compares, in a precise way, this new concept of space to the classical notion of topological space. This comparison takes the form of a two-fold extension of the idea of space.

Firstly, in classical topology a space is a set X equipped with a topology of open sets $\mathcal{O}(X) \subseteq \mathcal{P}(X)$. $\mathcal{O}(X)$ is required to be closed under arbitrary unions, and finite intersections. At the first level of extension, we replace the lattice of open sets by an arbitrary complete lattice satisfying the distributive law:

$$u \wedge (\bigvee_{i \in I} u_i) = \bigvee_{i \in I} (u \wedge u_i).$$

Such a lattice is called a locale. The concept of sheaf on a locale is clear, and gives rise to a corresponding topos. Morphisms of locales preserve suprema and finite infima - like the inverse image of open sets along a continuous map. The category of (extended) spaces and continuous maps is then defined to be the dual of the category of locales. This idea has a fairly long history in the literature, going back at least to Wallman (1938) [20]. For an exhaustive bibliography, see Johnstone [14]. It seems fair to say, however, that these authors were primarily intrigued by the possibility of doing "topology without points" (Papert 1964 [17]) rather than being forced to make the extension as one is in topos theory. We try here to lay a systematic foundation for this study, giving special attention to the concept of an open mapping of spaces, a notion we need for the proof of our principal theorem.

Secondly, we find that the difference between a general topos and sheaves on our new notion of space resides precisely in the possibility of action by a spatial groupoid. That is, if $G_1 \rightrightarrows G_0$ is a groupoid in the category of (extended) spaces, then we prove that the general notion of topos is captured by considering sheaves on G_0 together with a continuous action by G_1. Results of this kind should be thought of as a general type of Galois theory (hence the title of the paper), extending Grothendieck's interpretation [9] of classical Galois theory, in which he shows that the étale topos of a field k, which is the classifying topos for the theory of separably closed extensions of k, is equivalent to the category of continuous G-sets, where G is the profinite Galois group of a particular separable closure of k. Indeed, this turns out to be a special case of another structure theorem: any connected atomic topos with a point is continuous G-sets for an appropriate spatial group G.

Our basic technique is descent theory for morphisms of topoi and locales. Locales are commutative ringlike objects: supremum plays the role of addition, whereas infimum is the product. The structure corresponding to abelian group is sup-lattice. Once this is recognized, it is clear that the notion of module will play a central role. In fact, our first descent theorem for modules is completely analogous to the usual descent theorems of commutative algebra. After treating descent, we establish various facts about locales, and interpret them geometrically in the dual category, i.e. the category of spaces. The principal result states that open surjections of spaces are effective descent morphisms for sheaves. Later we extend the notion of open map to morphisms of topoi, and again prove that open surjections are effective descent morphisms for sheaves. This, together with the observation that any topos can be spatially covered by such a morphism yields the representation theorems.

A powerful method for studying a morphism of topoi $f\colon \mathcal{E}_2 \to \mathcal{E}_1$ is to regard $\mathcal{E}_2$ as a Grothendieck topos relative to $\mathcal{E}_1$. That is, we treat $\mathcal{E}_2$ as a category of $\mathcal{E}_1$-valued sheaves on a site in $\mathcal{E}_1$. The feasibility of this approach is due to the fact that we know how to interpret such set-theoretical notions as "site" and "sheaf" in $\mathcal{E}_1$. The possibility of interpreting general set-theoretical notions in an arbitrary Grothendieck topos was discovered by Lawvere and Tierney in their attempt to give an elementary axiomatization of the concept of topos. This resulted in the theory of elementary topoi, which provided the set-theoretical aspect of topos theory, expressed in categorical terms. Thus, we take the point of view that an arbitrary, but fixed, (elementary) topos is our basic universe S of sets. We work in this universe as we would in naive set theory, except that, of course, we cannot use the axiom of choice, or the law of the excluded middle. This has been shown to be correct by various authors (e.g. Boileau-Joyal [7]). As an illustration of the effectiveness of this method, we might cite, for example, chapter VI, §3, where it is shown that if A is a locale in S, then the category of A-modules in S is equivalent to the category of sup-lattices in the topos of sheaves on A, a result that cannot even be stated without adopting this position.

CHAPTER I - SUP-LATTICES

Our primary interest in the first few chapters is the study of locales in an arbitrary elementary topos S. As we mentioned in the Introduction, S should be thought of as the basic universe of generalized sets, and its objects will, in fact, be called "sets". Of course, we allow ourselves the freedom of changing the universe at will, which possibility, as will be seen in the sequel, is an important aspect of the theory.

Recall that a locale is a partially ordered set A admitting arbitrary suprema, hence finite infima, for which the distributive law

$$x \wedge (\bigvee_{i \in I} x_i) = \bigvee_{i \in I} (x \wedge x_i)$$

holds. Regarding the supremum as a kind of addition, and the infimum as a multiplication, a locale is clearly a kind of commutative ring. Before studying these, we are thus led to study the simpler structure analogous to that of abelian group, namely sup-lattice.

1. Definitions and duality

A partially ordered set M, which admits suprema of arbitrary subsets is called a sup-lattice. A morphism $f: M \to N$ of sup-lattices is a supremum-preserving map. We denote the category of sup-lattices of S by $s\ell$.

The supremum of $S \subseteq M$ will be written $\bigvee S$, or $\sup S$. The suprema of a family $(x_i)_{i \in I}$ of elements of M will be written $\bigvee_{i \in I} x_i$.

There is a partial order on the set $\mathrm{Hom}(M,N)$ of sup-lattice morphisms from M to N: $f \leq g$ iff $\forall x \in M$, $f(x) \leq g(x)$. In fact, this partial order is itself a sup-lattice:

$$(\bigvee_{i \in I} f_i)(x) = \bigvee_{i \in I} f_i(x).$$

As is well-known, a sup-lattice M also admits arbitrary infima: the infimum of a subset $S \subseteq M$ is calculated as the supremum of the set of lower bounds of S. Thus, if we denote the opposite partial order on M by M^0, then M^0 is also a sup-lattice.

If $f: M \to N$ is a morphism of sup-lattices, then f has a (unique) right adjoint $f_*: N \to M$ satisfying

$$\frac{f(x) \leq y}{x \leq f_*(y)} .$$

Received by the editors June 20, 1983.

f_* is easily calculated as

$$f_*(y) = \bigvee\{x \in M \mid f(x) \leq y\}.$$

As a right adjoint, f_* must preserve infima, and so defines a morphism of sup-lattices $f^0: N^0 \to M^0$. Clearly, $(M^0)^0 = M$, $(f^0)^0 = f$, and $(fg)^0 = g^0 f^0$. Also, $f \leq g$, iff $g_* \leq f_*$ iff $f^0 \leq g^0$. Hence

$$\mathrm{Hom}(M,N) \simeq \mathrm{Hom}(N^0,M^0).$$

Thus, we have

Proposition 1. The contravariant functor

$$(\ )^0: s\ell \to s\ell$$

is a (strong) self-duality.

2. Limits and colimits

The calculation of limits in $s\ell$ is very easy: the product $\prod_{i\in I} M_i$ is the product over I of the sets M_i with the coordinatewise partial order; the equalizer of a pair of morphisms $f,g: M \to N$ is the subset $\{x \in M \mid f(x) = g(x)\}$ with the induced partial order. In short, limits in $s\ell$ are calculated in S.

It follows that monomorphisms in $s\ell$ are injective, i.e. monomorphisms in S. In fact, let $m: M \to N$ be a monomorphism of $s\ell$. We have $mm_* \leq 1$, and $1 \leq m_*m$. Hence, $m = m(m_*m)$, and so $1 = m_*m$. Thus, monomorphisms of $s\ell$ have retractions in S. By duality, epimorphisms have sections.

In general, the duality is extremely useful for the calculation of colimits in $s\ell$ - usually much more difficult than limits.

For coproducts, consider the product $\prod_{i\in I} M_i$ of a family of sup-lattices. The projections $p_i: \prod_{i\in I} M_i \to M_i$ are calculated pointwise, as are the infima in $\prod_{i\in I} M_i$. Thus, the p_i preserve infima, and therefore have left adjoints $\mu_i: M_i \to \prod_{i\in I} M_i$. When I is decidable, the μ_i are given by the classical formula: $\mu_i(x_i) = (x_j)_{j\in I}$ where $x_j = 0$ $j \neq i$, $x_i = x$. In the general case we have

Proposition 1. The product $\prod_{i\in I} M_i$, equipped with the left adjoints $\mu_i: M_i \to \prod_{i\in I} M_i$ to the projections p_i, is the coproduct $\coprod_{i\in I} M_i$. Moreover, for each $i \in I$ $p_i\mu_i = 1$.

Proof: It is enough to show that $\mu_i^0: \prod_{i\in I} M_i^0 \to M_i^0$ is the projection.

But $\mu_i^0 = (\mu_i)_* = p_i$. For $p_i\mu_i = 1$, it is enough to show that μ_i is injective. For finite I this is clear, and for arbitrary I it is a consequence of the following:

Proposition 2. Let $(M_i, \alpha_{ij})_{i \in I}$ be a directed system of sup-lattices. Suppose that for $i \leq j$, $\alpha_{ij}: M_i \to M_j$ is injective. Then, for every $j \in I$, the canonical morphism

$$\alpha_j : M_j \to \varinjlim_{i \in I} M_i$$

is injective.

Proof: By duality it is equivalent to prove that the morphisms

$$\alpha_j^0 : \varprojlim_{i \in I} M_i^0 \to M_j^0$$

are surjective. Write $\alpha_{ij}^0 = \beta_{ji}$, $\alpha_j^0 = \beta_j$, and let $x \in M_j^0$. For $i \geq j$, put $x_i = (\beta_{ij})_*(x)$. We will be done if we show that if $k \geq i \geq j$, then $\beta_{ki}(x_k) = x_i$.

$$\begin{aligned} x_k &= (\beta_{kj})_*(x) \\ &= (\beta_{ij}\beta_{ki})_*(x) \\ &= (\beta_{ki})_*(\beta_{ij})_*(x) \\ &= (\beta_{ki})_*(x_i), \end{aligned}$$

and $\beta_{ki} \circ (\beta_{ki})_* = 1$, since β_{ki} is surjective. Thus:

$$\beta_{ki}(x_k) = \beta_{ki} \circ (\beta_{ki})_*(x_i) = x_i.$$

3. Free sup-lattices

As is well-known, the free sup-lattice on a set X in S is the power set PX equipped with the singleton map $X \xrightarrow{\{\ \}} PX$. Every sup-lattice is a quotient of one of these.

Just as clearly, the free sup-lattice on a partially ordered set Q is

$$DQ \subseteq PQ,$$

the set of downward closed subsets of Q, with union as the supremum, and down-segment $\downarrow(\ ): Q \to DQ$ as the universal map.

Since $P1$ is the free sup-lattice on 1, we have

$$M \simeq \mathrm{Hom}(P1, M).$$

By duality,

$$M^0 \simeq \mathrm{Hom}(M, P1^0).$$

4. Sub and Quotient Lattices

As we know from our calculation of limits, all monomorphisms of $s\ell$ are injective, and the sub-objects of a sup-lattice M are the subsets $S \subseteq M$ closed under suprema.

For any morphism $f\colon M \to N$, the set-image of f, $f(M) \subseteq N$, is a subobject, since it is closed under suprema. Moreover, the composite $\alpha = ff_*\colon N \to N$ is a <u>coclosure operator</u> on N, i.e.

1) $x \leq y \Longrightarrow \alpha(x) \leq \alpha(y)$

2) $\alpha(x) \leq x$

3) $\alpha^2(x) = \alpha(x)$.

Also, $f(M) = N_\alpha = \{y \in N \mid \alpha(y) = y\}$. In fact, we have

<u>Proposition 1</u>. There is a natural, order preserving isomorphism between the set of subobjects of N and the set of coclosure operators on N.

<u>Proof</u>: To a coclosure operator $\alpha\colon N \to N$, we assign the subobject $N_\alpha = \{y \in N \mid \alpha(y) = y\}$. To a subobject $S \subseteq N$, we assign the coclosure operator $\alpha_S\colon N \to N$ defined by

$$\alpha_S(x) = \bigvee\{y \in S \mid y \leq x\}.$$

By duality, given $f\colon M \to N$, the composite $\beta = f_*f\colon M \to M$ is a closure operator on M, i.e.

1) $x \leq y \Longrightarrow \beta(x) \leq \beta(y)$

2) $x \leq \beta(x)$

3) $\beta^2(x) = \beta(x)$.

Also, the set

$$M_\beta = \{x \in M \mid \beta(x) = x\}$$

is canonically isomorphic to $f(M)$. We have

<u>Proposition 2</u>. The following are canonically isomorphic:

1) The set of all quotients of M.

2) The set of all subsets $Q \subseteq M$, which are closed under infima.

3) The set of all closure operators β on M.

Here, the isomorphism between 1) and 2) is order preserving, but those between 2) and 3), and 1) and 3), are order reversing.

<u>Proof</u>: Let us just remark, that given a subset $Q \subseteq M$ closed under

infima, we can define a reflection $r: M \to Q$ by

$$r(x) = \bigwedge\{y \in Q \mid x \leq y\}.$$

r is left adjoint to the inclusion $Q \subseteq M$, so it preserves suprema and is surjective, exhibiting Q as a quotient of M.

As the final construction in this section, we describe explicitly how to compute the quotient of a sup-lattice by the congruence relation generated by an arbitrary subset of the product. This, of course, also gives an explicit calculation of coequalizers, thus finishing the description of colimits begun in §2.

Proposition 3. Let $R \subseteq M \times M$ be an arbitrary subset. The quotient of M by the congruence relation generated by R coincides with the subset

$$Q = \{x \in M \mid \forall (z_1, z_2) \in R,\ z_1 \leq x \Longleftrightarrow z_2 \leq x\}.$$

Proof: Q is clearly closed under infima. Let $r: M \to Q$ be the reflection. For $(z_1, z_2) \in R$ we have:

$$\begin{aligned} r(z_1) &= \bigwedge\{y \in Q \mid z_1 \leq y\} \\ &= \bigwedge\{y \in Q \mid z_2 \leq y\} \\ &= r(z_2). \end{aligned}$$

Conversely, suppose $r': M \to Q'$ is some other reflection such that for $(z_1, z_2) \in R$ $r'(z_1) = r'(z_2)$. If $x \in Q'$ and $(z_1, z_2) \in R$,

$$\begin{aligned} z_1 \leq x &\Longleftrightarrow r'(z_1) \leq x \\ &\Longleftrightarrow r'(z_2) \leq x \\ &\Longleftrightarrow z_2 \leq x. \end{aligned}$$

Thus, $Q' \subseteq Q$ and we are done.

5. Tensor products

If M, N and L are sup-lattices, we say a map $f: M \times N \to L$ is a bimorphism if f preserves suprema in each variable separately:

$$f\left(\bigvee_{i \in I} x_i, y\right) = \bigvee_{i \in I} f(x_i, y)$$

$$f\left(x, \bigvee_{j \in J} y_j\right) = \bigvee_{j \in J} f(x, y_j).$$

The tensor product $M \otimes N$ is then the codomain of the universal bi-morphism

$$M \times N \to M \otimes N.$$

As usual, $M \otimes N$ can be constructed as a quotient of the free sup-lattice on $M \times N$ - $P(M \times N)$ -by the equivalence relation generated by the obvious set of conditions (Banaschewski-Nelson [3]). We have the familiar adjointness

$$\mathrm{Hom}(M\otimes N,L) \simeq \mathrm{Hom}(M,\mathrm{Hom}(N,L)).$$

Proposition 1. The identities

$$\mathrm{Hom}(M,N) \simeq (N^0\otimes M)^0$$
$$M\otimes N \simeq \mathrm{Hom}(M,N^0)^0$$

hold.

Proof:
$$\begin{aligned}\mathrm{Hom}(M,N) &\simeq \mathrm{Hom}(N^0,M^0)\\ &\simeq \mathrm{Hom}(N^0,\mathrm{Hom}(M,P1^0))\\ &\simeq \mathrm{Hom}(N^0\otimes M,P1^0)\\ &\simeq (N^0\otimes M)^0.\end{aligned}$$

The second identity is a reformulation of the first - note that it implies that $M\otimes-$ preserves the order relation on maps.
For a full discussion of categories with identities of the above type, see Barr [4].

Proposition 2. Let $M \in s\ell$, and $X,Y \in S$. Then

i) $PX\otimes M \simeq X\cdot M$ - the coproduct of X copies of M.

ii) $P1\otimes M \simeq M$

iii) $PX\otimes PY \simeq P(X\times Y)$.

Moreover, for any family $(M_i)_{i\in I}$ of sup-lattices, we have

$$M \otimes \prod_{i\in I} M_i \simeq \prod_{i\in I} M\otimes M_i.$$

Proof: Only the last assertion needs to be checked. We know that $\prod_{i\in I} M_i$ equipped with the $\mu_i\colon M_i \to \prod_{i\in I} M_i$ left adjoint to the projections $p_i\colon \prod_{i\in I} M_i \to M_i$ is the coproduct of the family $(M_i)_{i\in I}$. As a result,

$$M\otimes\mu_i\colon M\otimes M_i \to M\otimes \prod_{i\in I} M_i$$

expresses $M\otimes \prod_{i\in I} M_i$ as the coproduct of the family $(M\otimes M_i)_{i\in I}$. But since $M\otimes(\ )$ preserves the order relation on maps, it follows that the morphisms $M\otimes p_i$ are right adjoint to the $M\otimes\mu_i$. Thus,

$$M\otimes p_i\colon M\otimes \prod_{i\in I} M_i \to M\otimes M_i$$

expresses $M\otimes \prod_{i\in I} M_i$ as the product of the family $(M\otimes M_i)_{i\in I}$.

CHAPTER II - RINGS, MODULES, AND DESCENT

Another description of a locale is that it is a sup-lattice A, for which the binary infimum

$$\wedge: A\times A \to A$$

preserves suprema in each variable separately. It thus passes to the tensor product giving a multiplication

$$\wedge: A\otimes A \to A,$$

which is associative and commutative in the obvious sense. This, together with the unit 1 makes A a commutative monoid in the symmetric, monoidal closed category $s\ell$. In fact, the locales are exactly those commutative monoids A satisfying $\forall a \in A,\ a \leq 1$ and $a^2 = a$. (We prove this in the next chapter.)

Our purpose in this chapter is to exploit more fully the analogy between sup-lattices and abelian groups, commutative monoids and commutative rings. Thus, we introduce the notion of a <u>module</u> for a commutative monoid, and study their basic properties. These include the elementary facts about tensor products, change of rings, flatness, projectivity, and purity. Finally, we investigate in more detail the <u>theory of descent</u> for modules, ending with a complete characterization of the effective descent morphisms for commutative monoids. These results will be applied to <u>spaces</u> - the dual of locales - in chapter V. We could call this the "algebraic aspect of the concept of space", and, as we shall see, it is a fruitful point of view.

1. <u>Rings and modules</u>

A <u>commutative monoid</u> in $s\ell$ is a sup-lattice A equipped with a multiplication

$$m: A\otimes A \to A$$

and a unit

$$1: 1 \to A.$$

Writing $m(a\otimes b) = a\cdot b$, these should satisfy

$$a\cdot b = b\cdot a$$

$$a\cdot(b\cdot c) = (a\cdot b)\cdot c$$

$$1\cdot a = a.$$

A <u>morphism</u> $f: A \to B$ of commutative monoids is a morphism of sup-lattices satisfying

$$f(a \cdot b) = f(a) \cdot f(b)$$
$$f(1) = 1$$

If A is a commutative monoid, an <u>A-module</u> is a sup-lattice M equipped with an <u>action</u>

$$\theta : A \otimes M \to M,$$

written $\theta(a \otimes x) = ax$, which is associative and unitary:

$$a(bx) = (a \cdot b)x$$
$$1x = x.$$

A morphism $f: M \to N$ of A-modules is a morphism of sup-lattices, which respects the actions:

$$f(ax) = af(x).$$

The category of A-modules is denoted by $\mathrm{Mod}(A)$.

Notice that $P1$ is a commutative monoid, in fact the initial commutative monoid, and for any sup-lattice M, the canonical isomorphism $P1 \otimes M \underset{\simeq}{\to} M$ is a $P1$-module structure on M. Thus, we have:

<u>Proposition 1</u>. The category $s\ell$ is canonically equivalent to the category $\mathrm{Mod}(P1)$.

As always, a module structure $A \otimes M \to M$ is equivalent to a morphism of monoids

$$A \to \mathrm{Hom}(M,M).$$

But $\mathrm{Hom}(M,M) = \mathrm{Hom}(M^0,M^0)$, so giving an A-module structure on M is equivalent to giving one on M^0. We can make this more explicit. For each $a \in A$, the map $a(\ ): M \to M$ preserves suprema, and hence has a right adjoint, which we denote by $(\ )^a$. The A-module structure $A \times M^0 \to M^0$ is given by $(a,x) \longmapsto x^a$.

<u>Proposition 2</u>. $\mathrm{Mod}(A)$ is (strongly) self dual.

2. <u>Tensor product of modules</u>

Let M, N, L be A-modules. A mapping $M \times N \to L$ is a <u>bi-morphism</u> of A-modules if it is a morphism of A-modules in each variable separately. The <u>tensor-product</u> $M \underset{A}{\otimes} N$ is the codomain of the universal bi-morphism of A-modules

$$M \times N \to M \underset{A}{\otimes} N.$$

As usual, it can be constructed as the evident coequalizer in the category $s\ell$:

$$M \otimes A \otimes N \rightrightarrows M \otimes N \twoheadrightarrow M \underset{A}{\otimes} N.$$

We have the familiar relations given by:

Proposition 1.

i) $\mathrm{Hom}_A(M \underset{A}{\otimes} N, L) \simeq \mathrm{Hom}_A(M, \mathrm{Hom}_A(N,L))$

ii) $\mathrm{Hom}_A(A,M) \simeq M$

iii) $M^0 \simeq \mathrm{Hom}_A(M, A^0)$

iv) $\mathrm{Hom}_A(M,N) \simeq (N^0 \underset{A}{\otimes} M)^0$

v) $M \underset{A}{\otimes} N \simeq \mathrm{Hom}_A(M, N^0)^0$

vi) $M \underset{A}{\otimes} \prod_{i \in I} M_i \simeq \prod_{i \in I} M \underset{A}{\otimes} M_i$.

3. Change of rings

If $f\colon A \to B$ is a morphism of commutative monoids and M is a B-module, then M can be regarded as an A-module by setting

$$ax = f(a)x.$$

The A-module so obtained will be denoted, when necessary, by $(M)_f$, or by $f^*(M)$. It is the result of restricting the scalars along f, which is a functor

$$(\ )_f\colon \mathrm{Mod}(B) \to \mathrm{Mod}(A).$$

Proposition 1. $(\ )_f$ has both a left and a right adjoint, denoted by

$$f_!, f_*\colon \mathrm{Mod}(A) \to \mathrm{Mod}(B)$$

respectively.

Proof: For the left adjoint, B itself can be regarded as an A-module as above, and given an A-module N, we can form $B \underset{A}{\otimes} N$, which is a B-module in an obvious way

$$f_!(N) = B \underset{A}{\otimes} N.$$

The right adjoint is given by

$$f_*(N) = \mathrm{Hom}_A(B,N).$$

The standard Hom-$\otimes$ formulas for bi-modules give the result.

We have the following (minor) applications. First, consider a diagram

$$\begin{array}{ccc} A & \xrightarrow{f} & B \\ {\scriptstyle g}\downarrow & & \\ C & & \end{array}$$

of commutative monoids. B and C both become A-modules as above, and we can form $C \otimes_A B$. It is a commutative monoid under the multiplication given by:

$$(c_1 \otimes b_1) \cdot (c_2 \otimes b_2) = (c_1 \cdot c_2) \otimes (b_1 \cdot b_2).$$

Moreover, we have morphisms

$$\mu_1 : C \to C \otimes_A B$$

$$\mu_2 : B \to C \otimes_A B$$

of commutative monoids given by $\mu_1(c) = c \otimes 1$, $\mu_2(b) = 1 \otimes b$, and these make

$$\begin{array}{ccc} A & \xrightarrow{f} & B \\ {\scriptstyle g}\downarrow & & \downarrow{\scriptstyle \mu_2} \\ C & \xrightarrow[\mu_1]{} & C \otimes_A B \end{array}$$

a pushout of commutative monoids in the usual way.

Secondly, for any commutative monoid A, we have the canonical morphism of commutative monoids

$$\mu : P1 \to A.$$

If N is an A-module, $(N)_\mu$ is just N regarded as a $P1$-module i.e. as a sup-lattice - $(\ )_\mu$ is the forgetful functor. It has left and right adjoints

$$s\ell \underset{\mathrm{Hom}(A,-)}{\overset{A\otimes -}{\rightrightarrows}} \mathrm{Mod}(A).$$

As a result, limits and colimits of A-modules are calculated in $s\ell$.

For $X \in S$, we have the <u>free module on</u> X, which is just

$$PX \otimes A \simeq X \cdot A \simeq A^X.$$

Every module is the quotient of a free module.

4. Flatness, projectivity, and purity

Definition. $M \in \mathrm{Mod}(A)$ is flat if $M\otimes_A -$ preserves monomorphisms. M is projective if $\mathrm{Hom}_A(M,-)$ preserves epimorphisms.

Proposition 1. The following are equivalent for an A-module A:

i) M is a retract of a free module
ii) M is projective
iii) M is flat.

Proof: i) $\Longrightarrow$ ii): $\mathrm{Hom}_A(X\cdot A, L) \simeq L^X$ preserves epimorphisms because they split in S. Any retract of $X\cdot A$ has the same property. ii) $\Longrightarrow$ i) is standard.
ii) $\Longleftrightarrow$ iii): $N' \overset{m}{\rightarrowtail} N$ is monic iff $N^0 \xrightarrow{m^0} N'^0$ is epic, so

$$\begin{array}{ccc} \mathrm{Hom}_A(M,N^0) & \xrightarrow{\mathrm{Hom}_A(M,m^0)} & \mathrm{Hom}_A(M,N'^0) \\ \downarrow \simeq & & \downarrow \simeq \\ (M\otimes_A N)^0 & \xrightarrow{(M\otimes_A m)^0} & (M\otimes_A N')^0 \end{array} .$$

M is projective iff $\mathrm{Hom}_A(M,m^0)$ is epic, iff $M\otimes_A m$ is monic, iff M is flat.
Let $M^* = \mathrm{Hom}_A(M,A)$.

Proposition 2. M is a retract of a free module iff $M^*\otimes_A -$ is left adjoint to $M\otimes_A -$.

Proof: Clearly, if $M^*\otimes_A -$ is left adjoint to $M\otimes_A -$, then the latter preserves monomorphisms and the result follows from proposition 1. So, suppose we have

$$X\cdot A \underset{s}{\overset{q}{\rightleftarrows}} M \qquad qs = 1.$$

For any A-modules N and L we have a transformation

$$L\otimes_A N \xrightarrow{\phi_N} \mathrm{Hom}_A(N^*,L)$$

natural in N (and L). It is the adjoint of the obvious map

$$L\otimes_A N\otimes_A N^* \to L\otimes_A A \simeq L.$$

If $N = X\cdot A$ we have

$$L \underset{A}{\otimes} X\cdot A \xrightarrow{\phi_{X\cdot A}} \mathrm{Hom}_A((X\cdot A)^*, L).$$

But $(X\cdot A)^* = \mathrm{Hom}_A(X\cdot A, A) \simeq A^X \simeq X\cdot A$.
Thus, $\phi_{X\cdot A}$ is equivalent to the canonical map

$$X\cdot L \to L^X,$$

which is an isomorphism. By the usual naturality argument, the property "ϕ_N is an isomorphism" is stable under retracts, and we are done.

<u>Definition</u>. A morphism $m: N' \to N$ of A-modules is <u>pure</u> iff for any A-module M,

$$M \underset{A}{\otimes} m: M \underset{A}{\otimes} N' \to M \underset{A}{\otimes} N$$

is monic.

Note that, taking $M = A$, a pure morphism is itself monic, but we can prove much more:

<u>Proposition 3</u>. Pure morphisms are retracts.

<u>Proof</u>: Suppose $m: N' \to N$ is pure, and $M \in \mathrm{Mod}(A)$. Consider

$$\begin{array}{ccc} \mathrm{Hom}_A(N,M) & \xrightarrow{\mathrm{Hom}_A(m,M)} & \mathrm{Hom}_A(N',M) \\ \simeq\downarrow & & \downarrow\simeq \\ (M^0 \underset{A}{\otimes} N)^0 & \xrightarrow{(M^0 \underset{A}{\otimes} m)^0} & (M^0 \underset{A}{\otimes} N')^0 \end{array}.$$

Since m is pure, $M^0 \underset{A}{\otimes} m$ is monic, and $\mathrm{Hom}_A(m,M)$ is epic. Taking $M = N'$ produces a retraction for m.

5. <u>Descent theory for modules</u>

We begin by describing the descent problem. Very roughly, it is the following: given a morphism $f: A \to B$ of commutative monoids, and a B-module M, when does there exist an A-module N such that $M \simeq B \underset{A}{\otimes} N$? There are certain, well-known, necessary conditions for this. (See, for example, Artin [1].) To explain these, consider

$$\begin{array}{ccccccc} & & M & & & & \\ A & \xrightarrow{f} & B & \underset{\mu_1}{\overset{\mu_2}{\rightrightarrows}} & B\underset{A}{\otimes}B & \underset{\mu_{12}}{\overset{\mu_{23}}{\underset{\mu_{13}}{\Rrightarrow}}} & B\underset{A}{\otimes}B\underset{A}{\otimes}B \\ & & & & \downarrow m & & \\ & & & & B & & \end{array}$$

where the subscripts on the μ's denote the position in the codomain tensor product in which the μ injects its domain. m is the codiagonal, i.e. the multiplication on B. Now if $M \simeq B\underset{A}{\otimes}N = f!(N)$, we will have an (iso)morphism

$$\phi\colon \mu_1!(M) \to \mu_2!(M)$$

of $B\underset{A}{\otimes}B$-modules (since $\mu_1 f = \mu_2 f$) such that

$$m!(\phi) = 1$$

and

$$\mu_{23}!(\phi)\cdot\mu_{12}!(\phi) = \mu_{13}!(\phi).$$

For an arbitrary M, such a structure ϕ is called <u>descent data on</u> M. The latter identity is called the <u>cocycle condition</u>, and together with $m!(\phi) = 1$ it implies ϕ is an isomorphism. On the other hand, if ϕ is an isomorphism satisfying the cocycle condition, then $m!(\phi) = 1$ follows. We may now ask: for which morphisms f are these necessary conditions (descent data) always sufficient? Let us make this question more precise.

First, sorting out a number of isomorphisms, descent data on M amounts to a morphism

$$\phi\colon M\underset{A}{\otimes}B \to B\underset{A}{\otimes}M$$

of $B\otimes B$-modules (with the obvious structures) such that

(1)
$$\begin{array}{ccc} M\underset{A}{\otimes}B & \xrightarrow{\ \phi\ } & B\underset{A}{\otimes}M \\ {\scriptstyle \tau_{MB}}\downarrow & & \downarrow{\scriptstyle \beta} \\ B\underset{A}{\otimes}M & \xrightarrow[\ \beta\ ]{} & M \end{array}$$

and

$$(2)\qquad \begin{array}{ccccc} & & M\underset{A}{\otimes}B\underset{A}{\otimes}B & & \\ & \overset{1\otimes\tau_{BB}}{\swarrow} & & \overset{\phi\otimes 1}{\searrow} & \\ M\underset{A}{\otimes}B\underset{A}{\otimes}B & & & & B\underset{A}{\otimes}M\underset{A}{\otimes}B \\ \downarrow{\scriptstyle \phi\otimes 1} & & & & \downarrow{\scriptstyle 1\otimes\phi} \\ B\underset{A}{\otimes}M\underset{A}{\otimes}B & & \xrightarrow[1\otimes\tau_{MB}]{} & & B\underset{A}{\otimes}B\underset{A}{\otimes}M \end{array}$$

commute, where β denotes the B-module structure of M, and a τ with two subscripts indicates the symmetry isomorphism on the $\otimes$-product of the subscripts. The canonical ϕ for a B-module of the form $B\underset{A}{\otimes}N$ is

$$1\otimes\tau_{NB}\colon (B\underset{A}{\otimes}N)\underset{A}{\otimes}B \to B\underset{A}{\otimes}(B\underset{A}{\otimes}N).$$

By Des(f) we mean the category whose objects are pairs (M,ϕ), where M is a B-module, and ϕ is descent data on M. Morphisms are morphisms of B-modules compatible in the obvious sense with the descent data. We obtain a diagram

$$\begin{array}{ccc} \mathrm{Mod}(A) & \xrightarrow{\ \Phi\ } & \mathrm{Des}(f) \\ & {\scriptstyle B\underset{A}{\otimes}-}\searrow\nwarrow{\scriptstyle (\ )_f} & \downarrow{\scriptstyle U} \\ & & \mathrm{Mod}(B), \end{array}$$

where $U(M,\phi) = M$, and $\Phi(N) = (B\underset{A}{\otimes}N,\ 1\otimes\tau_{NB})$.

<u>Definition</u>. f is called a <u>descent morphism</u> if Φ is full and faithful, and an <u>effective descent morphism</u> if Φ is an equivalence.

Our precise problem here is: what are the effective descent morphisms of commutative monoids? The answer is the same as in the case of commutative algebra in the category of classical sets: f is an effective descent morphism iff f is pure as a morphism of A-modules. We remark that, refining the techniques of the present section, we have extended this to the case of commutative rings and modules in an arbitrary topos S, though we will not treat it here since it does not really fit our program. In the present case - of $s\ell$ instead of abelian groups - we will see that the proof of this result, even for an arbitrary S, is considerably easier than the classical one, due to the fact, established in §4, that pure morphisms are retracts. We begin by reformulating the descent conditions.

We observe that descent data is given by a morphism $\phi: \mu_1!(M) \to \mu_2!(M)$. But by the adjointness of §3, such a morphism is equivalent to a morphism

$$\theta: M \to \mu_1^*\mu_2!(M).$$

Calculating, we see that $\mu_2!(M) = (B \underset{A}{\otimes} B) \underset{B}{\otimes} M \simeq B \underset{A}{\otimes} M$ as $B \underset{A}{\otimes} B$-modules, and $\mu_1^*\mu_2!(M)$ is thus $B \underset{A}{\otimes} M$, where the B-structure is given by multiplication in the first factor. But this is just $f!f^*(M)$, so the canonical natural transformation $f!f^*(M) \to \mu_1^*\mu_2!(M)$, which always exists, is an equivalence, and descent data is determined by a morphism

$$\theta: M \to B \underset{A}{\otimes} M.$$

In fact, θ is the composite

$$M \to M \underset{A}{\otimes} B \xrightarrow{\phi} B \underset{A}{\otimes} M,$$

where the first morphism is $M \simeq M \underset{A}{\otimes} A \xrightarrow{1 \otimes f} M \underset{A}{\otimes} B$ given by $x \longmapsto x \otimes 1$. In general, for an A-module N, we write ηN for the map $N \simeq A \underset{A}{\otimes} N \xrightarrow{f \otimes 1} B \underset{A}{\otimes} N$ given by $y \longmapsto 1 \otimes y$, as it is the unit of the adjunction $f! \dashv f^*$.

To complete the translation, we should verify that ϕ is descent data - i.e. makes (1) and (2) commute, iff θ makes

(1')
$$\begin{array}{ccc} M & \xrightarrow{\theta} & B \underset{A}{\otimes} M \\ & \diagdown\!\!\diagdown & \big\downarrow \beta \\ & & M \end{array}$$

and

(2')
$$\begin{array}{ccc} M & \xrightarrow{\theta} & B \underset{A}{\otimes} M \\ \theta \big\downarrow & & \big\downarrow 1 \otimes \theta \\ B \underset{A}{\otimes} M & \xrightarrow[1 \otimes \eta M]{} & B \underset{A}{\otimes} B \underset{A}{\otimes} M \end{array}$$

commute. Further, we should show that a morphism of B-modules is compatible with the ϕ's iff with the θ's. This is lengthy, but straightforward, and we omit the details.

We must point out that this translation is a special case of a much more general result due to Beck (unpublished). He shows that for a general bifibration - a situation such as ours where over every morphism f of a "base" category there lies an adjoint pair $f! \dashv f^*$ - if the

Beck-Chevalley condition is satisfied - meaning the natural transformations $f_!f^* \to \mu_1{}^*\mu_{2!}$ are all equivalences - then descent data (ϕ as above) is equivalent to an $f_!f^*$-coalgebra structure (θ).

Writing Des(f) again for the category of pairs (M,θ), we obtain a diagram

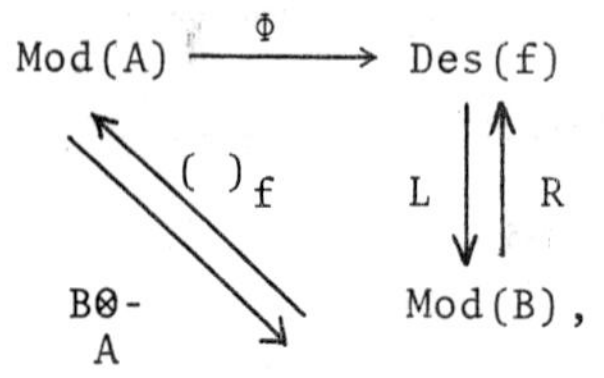

where $L(M,\theta) = M$, $R(M) = (B\otimes_A M,\ 1\otimes\eta M)$ and L is left adjoint to R. $\Phi(N) = (B\otimes_A N,\ 1\otimes\eta N)$.

We start attacking the descent problem by constructing a right adjoint $\hat{\Phi}$ to Φ. $\hat{\Phi}$ is defined by requiring that

$$\hat{\Phi}(M,\theta) \xrightarrow{e} M \underset{\eta M}{\overset{\theta}{\rightrightarrows}} B\otimes_A M$$

be an equalizer of A-modules for each $(M,\theta) \in \mathrm{Des}(f)$. We indicate the unit and counit of the adjunction $\Phi \dashv \hat{\Phi}$, and leave details to the reader.

First, note that $\hat{\Phi}\Phi(N)$ is the equalizer in the bottom row in

(a)

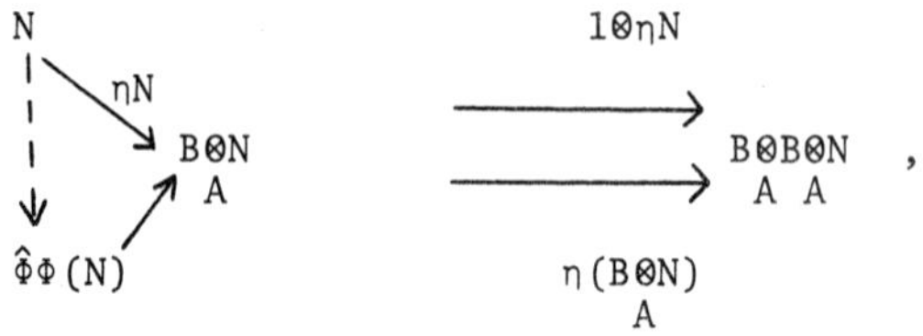

and the canonical filler for the dotted arrow is the unit. It exists because ηN equalizes the two right hand morphisms.

Next, the canonical filler for the dotted arrow in

(b)

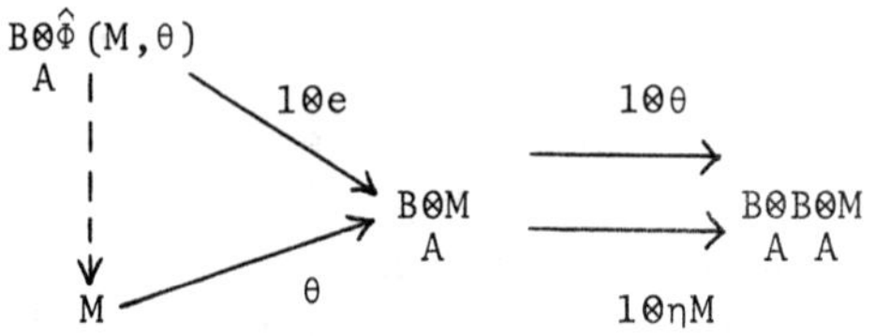

is L of the counit $\Phi\hat{\Phi}(M,\theta) \to (M,\theta)$. It is the adjoint to the given morphism $\hat{\Phi}(M,\theta) \xrightarrow{e} (M)_f$.

Considering diagrams (a) and (b), we obtain immediately two propositions. These are the basis of all descent theorems.

<u>Proposition 1</u>. Φ is full and faithful, i.e. f is a descent morphism, iff

$$N \xrightarrow{\eta N} B\underset{A}{\otimes}N \;\underset{\eta(B\underset{A}{\otimes}N)}{\overset{1\otimes\eta N}{\rightrightarrows}}\; B\underset{A}{\otimes}B\underset{A}{\otimes}N$$

is an equalizer for each A-module N.

<u>Proposition 2</u>. A descent morphism f is effective, i.e. Φ is an equivalence, iff $B\underset{A}{\otimes}-$ preserves the equalizer

$$\hat{\theta}(M,\theta) \xrightarrow{e} M \;\underset{\eta M}{\overset{\theta}{\rightrightarrows}}\; B\underset{A}{\otimes}M$$

for each $(M,\theta) \in \mathrm{Des}(f)$.

Let us now investigate these conditions more closely.

<u>Proposition 3</u>.

$$N \xrightarrow{\eta N} B\underset{A}{\otimes}N \;\underset{\eta(B\underset{A}{\otimes}N)}{\overset{1\otimes\eta N}{\rightrightarrows}}\; B\underset{A}{\otimes}B\underset{A}{\otimes}N$$

is an equalizer for each A-module N iff $\eta N\colon N \to B\underset{A}{\otimes}N$ is monic for each N.

<u>Proof</u>: One direction is clear, so suppose each ηN is monic. Since each monomorphism of $s\ell$ is an equalizer, it follows that there is an equalizer diagram

$$N \xrightarrow{\eta N} B\underset{A}{\otimes}N \;\underset{g}{\overset{k}{\rightrightarrows}}\; L$$

for some k, g and L in $\mathrm{Mod}(A)$. To show that the diagram of the proposition is an equalizer it is enough, since it equalizes, to show that if a morphism $h\colon M \to B\underset{A}{\otimes}N$ of A-modules equalizes $1\otimes\eta N$ and $\eta(B\underset{A}{\otimes}N)$, it must equalize k and g, for then it factors uniquely through ηN. For this, consider the diagram

$$\begin{array}{ccc}
M & & \\
\downarrow{\scriptstyle h} & & \\
B\otimes_A N & \underset{\eta(B\otimes_A N)}{\overset{1\otimes\eta N}{\rightrightarrows}} & B\otimes_A B\otimes_A N \\
{\scriptstyle k}\downarrow\downarrow{\scriptstyle g} & & {\scriptstyle 1\otimes k}\downarrow\downarrow{\scriptstyle 1\otimes g} \\
L & \xrightarrow[\eta L]{} & B\otimes_A L
\end{array}$$

We have:

$1\otimes k\circ\eta(B\otimes_A N)\circ h = \eta L\circ k\circ h$ and $1\otimes g\circ\eta(B\otimes_A N)\circ h = \eta L\circ g\circ h$ by naturality of η. But $\eta(B\otimes_A N)\circ h = 1\otimes_A\eta N\circ h$ by assumption, and $1\otimes k\circ 1\otimes\eta N = 1\otimes g\circ 1\otimes\eta N$ since $k\circ\eta N = g\circ\eta N$. Thus, $\eta L\circ k\circ h = \eta L\circ g\circ h$. ηL is monic, so $k\circ h = g\circ h$ as desired.

Theorem 1. $f\colon A \to B$ is a descent morphism iff f is pure as a morphism of A-modules.

Proof: This follows immediately from Proposition 1, Proposition 3, and the definition of purity, since N is

$$N \simeq A\otimes_A N \xrightarrow{f\otimes 1} B\otimes_A N.$$

Theorem 2. If $f\colon A \to B$ has an A-linear retraction, then f is an effective descent morphism.

Proof: We first need a bit of terminology. A diagram

$$M' \underset{e}{\overset{r}{\leftrightarrows}} M \underset{f_2}{\overset{f_1}{\rightrightarrows}} M'' \qquad s\colon M'' \to M$$

is called a split equalizer if

$$f_1e = f_2e$$
$$re = 1$$
$$sf_2 = 1$$
$$sf_1 = er\ .$$

It follows easily that in a split equalizer,

$$M' \xrightarrow[e]{} M \underset{f_2}{\overset{f_1}{\rightrightarrows}} M''$$

is an equalizer, one which, by its equational nature, is preserved by any functor.

Now let $r: B \to A$ be a morphism of A-modules such that $rf = 1$. r provides a natural (in N) retraction $rN = r\otimes_A N: B\otimes_A N \to A\otimes_A N \simeq N$ for each ηN, so, for each N,

$$\begin{array}{ccccc} & \xrightarrow{\eta N} & & \xrightarrow{1\otimes\eta N} & \\ N & & B\otimes_A N & \xrightarrow[\eta(B\otimes_A N)]{} & B\otimes_A B\otimes_A N \\ & \xleftarrow[rN]{} & & \xleftarrow[1\otimes rN]{} & \end{array}$$

is a split equalizer of A-modules. Thus, f is a descent morphism, which, in any case, is clear from Theorem 1. In addition, however, let $(M,\theta) \in \mathrm{Des}(f)$ and consider the diagram

$$\begin{array}{ccccc} M & \xrightarrow{\theta} & B\otimes_A M & \underset{1\otimes\eta M}{\overset{1\otimes\theta}{\rightrightarrows}} & B\otimes_A B\otimes_A M \\ {\scriptstyle r}\,\big\downarrow & & {\scriptstyle rM}\big\downarrow\big\uparrow{\scriptstyle \eta M} & & {\scriptstyle r(B\otimes_A M)}\big\downarrow\big\uparrow{\scriptstyle \eta(B\otimes_A M)} \\ \hat{\Phi}(M,\theta) & \xrightarrow{e} & M & \underset{\eta M}{\overset{\theta}{\rightrightarrows}} & B\otimes_A M \end{array}$$

By naturality, both right hand squares (with the r's) commute, providing a unique $r: M \to \hat{\Phi}(M,\theta)$ such that $er = rM\circ\theta$. But then the bottom line becomes a split equalizer of A-modules by means of $r: M \to \hat{\Phi}(M,\theta)$ and $rM: B\otimes_A M \to M$. Therefore, $B\otimes_A(\ )$ applied to it remains an equalizer, and Φ is an equivalence by Proposition 2.

By Proposition 3 §4, pure morphisms of A-modules have A-linear retractions, so we have

<u>Theorem 3</u>. $f: A \to B$ is an effective descent morphism for modules iff f is pure as a morphism of A-modules.

We make a few remarks now on descent of structure, since we will need them in the following chapters. One basic fact underlies all the discussion. Namely, let $g: C \to D$ be a morphism of commutative monoids. Then

$$\begin{aligned} g!M\otimes g!N &= (D\otimes_C M)\otimes_D(D\otimes_C N) \\ &\simeq D\otimes_C(M\otimes_C N) = g!(M\otimes_C N), \end{aligned}$$

so $g!$ preserves $\otimes$-product. As a consequence, consider

$$A \xrightarrow{f} B \underset{\mu_1}{\overset{\mu_2}{\rightrightarrows}} B\underset{A}{\otimes}B \; \overrightarrow{\underset{\longrightarrow}{\longrightarrow}} \cdots ,$$

where f is a descent morphism, and suppose M is a B-module with effective descent data $\theta: M\underset{A}{\otimes}B \to B\underset{A}{\otimes}M$; $M \simeq f!(N)$. Moreover, suppose M is, say, a B-algebra (a monoid in $\mathrm{Mod}(B)$). Thus, we have an associative multiplication m:

$$\begin{array}{ccc} M\underset{B}{\otimes}M & \xrightarrow{m} & M \\ \simeq\big\downarrow & & \big\downarrow\simeq \\ f!(N\underset{A}{\otimes}N) \simeq f!N\underset{B}{\otimes}f!N & \longrightarrow & f!N \end{array}$$

with unit 1:

$$\begin{array}{ccc} B & \xrightarrow{1} & M \\ \simeq\big\downarrow & & \big\downarrow\simeq \\ f!A & \longrightarrow & f!N \end{array} .$$

Consider

$$\begin{array}{ccc} \mu_1!(M\underset{B}{\otimes}M) \simeq \mu_1!M \underset{B\underset{A}{\otimes}B}{\otimes} \mu_1!M & \xrightarrow{\mu_1!(m)} & \mu_1!M \\ \big\downarrow \theta\otimes\theta & & \big\downarrow \theta \\ \mu_2!(M\underset{B}{\otimes}M) \simeq \mu_2!M \underset{B\underset{A}{\otimes}B}{\otimes} \mu_2!M & \xrightarrow{\mu_2!(m)} & \mu_2!M \end{array} .$$

As we see, $\mu_1!M$ and $\mu_2!M$ are $B\underset{A}{\otimes}B$-algebras, $M\underset{B}{\otimes}M$ has effective descent data $\theta\otimes\theta$ on it, and if θ is an <u>algebra morphism</u>, then m is compatible and hence descends uniquely. Similar remarks apply to the unit. In the same way, any further axioms satisfied by m and 1 expressible by commutative diagrams involving $\otimes$-product also descend - e.g. commutativity, etc., etc.. Later we will have to consider descent of properties, which are not quite of this type, but we leave those until we need them.

CHAPTER III - LOCALES

As the reader will recall, a locale A is a commutative monoid in $s\ell$ such that $a \cdot b = a \wedge b$. The fact that the binary infimum preserves suprema in each variable is equivalent to the existence, for each $a \in A$, of a right adjoint $a \to (\ ): A \to A$ to the mapping $(\ ) \wedge a: A \to A$ i.e.

$$\frac{b \wedge a \leq c}{b \leq a \to c} .$$

Thus, a locale is a complete Heyting algebra. However, a morphism of locales is just a morphism of commutative monoids $f: A \to B$:

$$f(a \wedge b) = f(a) \wedge f(b)$$
$$f(1) = 1.$$

The category of locales is denoted by Loc.

1. Locales and commutative monoids

For later applications of the descent theorems, we need to characterize locales among the commutative monoids of $s\ell$.

Proposition 1. A commutative monoid A is a locale iff A satisfies $\forall a \in A$, $a \leq 1$ and $a^2 = a$.

Proof: The conditions are clearly necessary, so suppose they are satisfied in A. Then

$$a \cdot b \leq a \cdot 1 = a$$
$$a \cdot b \leq 1 \cdot b = b,$$

since multiplication is order preserving. On the other hand, if $c \leq a$ and $c \leq b$, then $c = c \cdot c \leq c \cdot b \leq a \cdot b$, so $a \cdot b = a \wedge b$.

2. Limits and colimits

Limits in Loc are easy to calculate: just take the limit of the underlying sets.

The coproduct of two locales A and B is their coproduct as commutative monoids: $A \otimes B$.

Let $(A_i, \alpha_{ij})_{i \in I}$ be a directed system of locales. Since I is directed, $\Delta: I \to I \times I$ is cofinal, and in $s\ell$ we have:

$$\varinjlim_I A_i \otimes \varinjlim_I A_j \simeq \varinjlim_{I\times I} (A_i \otimes A_j) \simeq \varinjlim_I (A_i \otimes A_i).$$

Thus, the colimit $\varinjlim_I A_i$ in $s\ell$ is also the colimit in Loc, since the infimum is the colimit over I of the infima

$$A_i \otimes A_i \xrightarrow{\wedge} A_i .$$

This is, of course, also true for a direct system of arbitrary commutative monoids. In any case, Proposition 2 chapter I §2 yields

Proposition 1. Let $(A_i, \alpha_{ij})_{i\in I}$ be a direct system of locales such that for $i \leq j$, $\alpha_{ij}: A_i \to A_j$ is injective. Then, for any $i \in I$ the canonical map $A_i \to \varinjlim_I A_i$ is injective.

For an arbitrary family of locales, the coproduct $\coprod_{i\in I} A_i$ is

$$\varinjlim_{\{i_1,\ldots,i_n\}\subseteq I} A_{i_1} \otimes \ldots \otimes A_{i_n} .$$

3. The free locale

The free locale $L(X)$ on a set $X \in S$ can be constructed in two steps.

First, take the free inf-semi-lattice on X: this is $K(X)^0$, where

$$K(X) \subseteq P(X)$$

is the set of all (Kuratowski) finite subsets of X. This is the smallest subset of $P(X)$ containing the singletons, the empty set, and closed under binary union. The universal map $X \to K(X)^0$ is the singleton mapping.

For an inf semi-lattice Z, the free locale on Z is the free sup-lattice on the partial order Z - i.e. the set $\mathcal{D}(Z)$ of downward closed subsets of Z from Chapter I §3. The universal map $Z \to \mathcal{D}(Z)$ is the down-segment mapping.

4. Local operators and quotients

If $f: A \to B$ is a morphism of locales, the composite $\ell = f_* f: A \to A$ is a closure operator preserving finite infima since both f and f_* do.

Proposition 1. The following conditions on a closure operator $\ell: A \to A$ are equivalent

i) $\ell(a \wedge b) = \ell(a) \wedge \ell(b)$

ii) $a \to b \leq \ell(a) \to \ell(b)$

iii) $\ell(a \to \ell(b)) = a \to \ell(b)$.

Proof: The first condition is equivalent to the principle

$$\frac{a \wedge b \leq c}{\ell(a) \wedge \ell(b) \leq \ell(c)},$$

the second to

$$\frac{a \wedge b \leq c}{a \wedge \ell(b) \leq \ell(c)}$$

and the third to

$$\frac{a \wedge b \leq c}{a \wedge \ell(b) \leq c} \quad \ell(c) = c$$

These three principles are clearly equivalent.

Definition. A local operator $\ell: A \to A$ is a closure operator satisfying any one of the three previous equivalent conditions.

Proposition 2. For a locale A, the following are canonically isomorphic.

i) The set of all quotients of A.

ii) The set of subsets $Q \subseteq A$ closed under infima and such that if $a \in A$ and $x \in Q$, then $a \to x \in Q$.

iii) The set of local operators on A.

Proof: If $q: A \to Q$ is a quotient locale, then $\ell = q_* q$ is a local operator, and $Q \simeq \{x \in A | \ell(x) = x\}$. If $a \in A$ and $x \in Q$, then

$$\ell(a \to x) = \ell(a \to \ell(x)) = a \to \ell(x) = a \to x,$$

so Q satisfies ii). This is true for an arbitrary local operator. Now suppose Q is any subset of A satisfying ii). If $x, y \in Q$ then $x \to y \in Q$, so Q is a locale. We must show that the associated reflection $q: A \to Q$ preserves finite infima. This is equivalent with saying that the associated closure operator ℓ is a local operator. But the identity $\ell(a \to \ell(b)) = a \to \ell(b)$ means exactly that $a \to \ell(b)$ is closed - i.e. $a \to x$ is closed if x is.

The sup-lattice $P1$ is a locale, in fact it is the initial locale. It enjoys a special property:

Proposition 3. Any closure operator $\ell: P1 \to P1$ is a local operator.

Proof: The mappings

$$(x,y) \longmapsto x \to y$$

$$(x,y) \longmapsto \ell(x) \to \ell(y)$$

are characteristic maps

$$\chi_{S_1}, \chi_{S_2} : P1 \times P1 \to P1$$

of subsets $S_1, S_2 \subseteq P1 \times P1$, and it is enough to check that $S_1 \subseteq S_2$, for then we have $x \to y \leq \ell(x) \to \ell(y)$. But for this it is enough to know that $x \leq y \Longrightarrow \ell(x) \leq \ell(y)$, which is true.

As we did for sup-lattices given a subset $R \subseteq A \times A$, we want to explicitly describe the quotient locale Q of A by the congruence relation on A generated by R. We say that R is inf-stable if given $a \in A$ and $(z_1, z_2) \in R$, we have $(a \wedge z_1, a \wedge z_2) \in R$. Replacing R by

$$A \wedge R = \{(a \wedge z_1, a \wedge z_2) \mid a \in A,\ (z_1, z_2) \in R\}$$

we may suppose R is inf-stable, for R and $A \wedge R$ generate the same congruence relation.

Proposition 4. Let A be a locale, and $R \subseteq A \times A$ an inf-stable subset. The quotient of A by the congruence relation generated by R is given by the subset

$$Q = \{x \in A \mid \forall (z_1, z_2) \in R,\ z_1 \leq x \Longleftrightarrow z_2 \leq x\}.$$

Proof: By Proposition 2, and Proposition 3 Chapter I §4, all we need to show is that if $a \in A$ and $x \in Q$, then $a \to x \in Q$. For this, suppose $(z_1, z_2) \in R$. Then

$$\frac{\dfrac{\dfrac{z_1 \leq a \to x}{a \wedge z_1 \leq x}}{a \wedge z_2 \leq x}}{z_2 \leq a \to x}$$

and we are done.

As an application of Proposition 4, we consider the notion of a Grothendieck topology, or rather a covering system on a partially ordered set P. Such a system is given by specifying, for each $x \in P$, a collection $\mathrm{Cov}(x) \subseteq P{\downarrow}(x)$ of covers of x - i.e. a cover $R \in \mathrm{Cov}(x)$ is a given subset of the down segment of x. Let $\hat{R}$ denote the downward closed subset of P generated by R: $z \in R$ iff $z \leq y$ for some $y \in R$. Then the condition that these covers should satisfy is: if $y \leq x$ and $R \in \mathrm{Cov}(x)$, then $\hat{R} \wedge {\downarrow}(y) \supseteq R'$, where $R' \in \mathrm{Cov}(y)$. This is not yet a topology on P, since we have not required that singletons be covers, or that the local axiom is satisfied. It is a system of generators for a topology.

Now consider the locale $\mathcal{D}(P)$ of downward closed subsets of P, and the subset of $\mathcal{D}(P) \times \mathcal{D}(P)$ consisting of pairs $(\hat{R}, {\downarrow}(x))$ where $x \in P$ and $R \in \mathrm{Cov}(x)$. Making this inf-stable, we obtain the set of pairs

$(\hat{R} \wedge S, \downarrow(x) \wedge S)$ where $S \in \mathcal{D}(P)$ and $R \in \mathrm{Cov}(x)$. By Proposition 4 the locale quotient Q of $\mathcal{D}(P)$ by the generated congruence relation is

$$Q = \{T \in \mathcal{D}(P) \mid \forall x \in P\ \forall R \in \mathrm{Cov}(x)\ \forall S \in \mathcal{D}(P)\ \hat{R} \wedge S \subseteq T \text{ iff } \downarrow(x) \wedge S \subseteq T\}.$$

An immediate calculation shows that this is the set

$$Q = \{T \in \mathcal{D}(P) \mid \forall x \in P\ \forall R \in \mathrm{Cov}(x)\ R \subseteq T \Longrightarrow x \in T\}.$$

Q, of course, gives Ω in the topos of sheaves for the generated topology on P.

5. The splitting locale

Let A be a locale, and $a \in A$. A complement for a is an element $a' \in A$, such that $a \wedge a' = 0$, $a \vee a' = 1$. For any $S \subseteq A$ we will write

$$h^S\colon A \to A_S$$

for the universal solution - if it exists - to the problem of adding a complement for each element of S. If $S = A$, we write $h\colon A \to A'$ for the solution.

Proposition 1. For any $S \subseteq A$, $h^S\colon A \to A_S$ exists, and is injective. In particular, $h\colon A \to A'$ is injective.

Proof: We first show the existence and injectivity of h^F where $F \subseteq A$ is finite. If $F = \{a_1,\ldots,a_n\}$, we can construct A_F by successively adding complements to a_1 up to a_n. Thus, the problem is reduced to the case $F = \{a\}$. Consider

$$[0,a] = \{b \in A \mid 0 \le b \le a\}$$
$$[a,1] = \{b \in A \mid a \le b \le 1\}.$$

Each of these is a locale, and there is a morphism of locales

$$h^a\colon A \to [0,a]\times[a,1]$$

given by $h^a(x) = (x \wedge a, x \vee a)$. Direct calculation shows that h^a is the universal solution to the problem of adding a complement to a. The injectivity of h^a is equivalent to the statement that $x \wedge a = y \wedge a$ and $x \vee a = y \vee a$ entails $x = y$, which is true.

Now let I be the directed set of all finite subsets of $S \subseteq A$. We have a directed system $(A_F)_{F \in I}$, and when $F \subseteq G$, the transition mapping $A_F \to A_G$ is injective, since $A_G = (A_F)_G$. Clearly,

$$A_S = \varinjlim_{F \in I} A_F.$$

The injectivity of $h^S: A \to A_S$ is a consequence of Proposition 1 §2.

Definition. The locale A' is called the splitting locale of A.

Proposition 2. Any element $x \in A'$ can be expressed as a supremum of elements of the form $a \wedge b'$, where $a \in A$, $b \in A$, and b' is the complement of b in A'.

Proof: The elements of A' of the form $a \wedge b'$ are closed under finite infima. Let B be the set of all suprema of these elements. B is a sublocale of A' and $A \subseteq B$. By the universal property, $B = A'$.

CHAPTER IV - SPACES

Here we study "the geometric aspect of the concept of Locale", by introducing the category of <u>spaces and continuous maps of</u> <u>S</u>. This, by definition, is the category dual to Loc:

$$Sp = Loc^{op}.$$

For $X \in Sp$, we write $\mathcal{O}(X) \in Loc$ for the corresponding locale, and we call $\mathcal{O}(X)$ the <u>lattice of open parts of</u> X. If $f: X \to Y$ is a continuous map of spaces, we write $f^{-}: \mathcal{O}(Y) \to \mathcal{O}(X)$ for the corresponding morphism of locales. For $f,g: X \to Y$, put $f \leq g$ iff $g^{-} \leq f^{-}$ i.e. for every $v \in \mathcal{O}(Y)$, $g^{-}(v) \leq f^{-}(v)$.

Much of the material of this chapter is simply dual to that of Chapter III, but it is nonetheless necessary for the development of a correct geometric intuition.

1. <u>Subspaces</u>

The concept of subspace is dual to the concept of quotient locale. Thus, the subspaces of X are in one to one correspondence with the local operators $\ell: \mathcal{O}(X) \to \mathcal{O}(X)$. If X_ℓ denotes the subspace corresponding to ℓ, we have $\mathcal{O}(X_\ell) = \mathcal{O}(X)_\ell$ - the lattice of ℓ-closed elements of $\mathcal{O}(X)$. A subspace of X is also called a <u>part</u> of X.

For any element $u \in \mathcal{O}(X)$, we have an associated <u>open subspace</u> $U \hookrightarrow X$: $\mathcal{O}(U) = \{v \in \mathcal{O}(X) | v \leq u\}$, and the quotient mapping $\mathcal{O}(X) \to \mathcal{O}(U)$ is $w \longmapsto w \wedge u$. The corresponding local operator is $u \to (\ ): \mathcal{O}(X) \to \mathcal{O}(X)$. $\mathcal{O}(U)$ is the quotient of $\mathcal{O}(X)$ by the congruence relation generated by the condition $u \equiv 1$.

Similarly, for any $u \in \mathcal{O}(X)$, we have an associated <u>closed</u> subspace $CU \hookrightarrow X$. $\mathcal{O}(CU) = \{v \in \mathcal{O}(X) | u \leq v\}$. The quotient mapping $\mathcal{O}(X) \to \mathcal{O}(CU)$ is $w \longmapsto w \vee u$, and the corresponding local operator is $u \vee (\ ): \mathcal{O}(X) \to \mathcal{O}(X)$. $\mathcal{O}(CU)$ is the quotient of $\mathcal{O}(X)$ by the congruence relation generated by the condition $u \equiv 0$.

<u>Definition</u>. Let $\ell_1, \ell_2: \mathcal{O}(X) \to \mathcal{O}(X)$ be local operators. We say ℓ_1 is <u>distributive</u> over ℓ_2 if we have $\ell_1\ell_2 \leq \ell_2\ell_1$.

<u>Proposition 1</u>. If ℓ_1 is distributive over ℓ_2, then $\ell = \ell_2\ell_1$ is a local operator, and X_ℓ is the intersection of X_{ℓ_1} and X_{ℓ_2}.

Proof: The fact that if $\ell_1\ell_2 \leq \ell_2\ell_1$ then $\ell_2\ell_1$ is a local operator is immediate. $X_\ell = X_{\ell_1} \cap X_{\ell_2}$ means ℓ is the sup of ℓ_1 and ℓ_2 in the partially ordered set of local operators, which is also clear.

Corollary: Let $\ell\colon \mathcal{O}(X) \to \mathcal{O}(X)$ be a local operator and $u \in \mathcal{O}(X)$. The intersection $U \wedge X_\ell$ is described by the local operator $u \to \ell(\ )$, and the intersection $CU \wedge X_\ell$ is described by the local operator $\ell(u \vee (\ ))$.

Proof: The distributivities $\ell(u \to x) \leq u \to \ell(x)$ (and $u \vee \ell(x) \leq \ell(u \vee x)$ are clear).

A subspace which is the intersection of an open subspace V and a closed subspace CU is called a locally closed subspace.

$$\mathcal{O}(V \wedge CU) = \{x \in \mathcal{O}(X) \mid u \wedge v \leq x \leq v\}.$$

Finally, let us remark that a continuous mapping $f\colon X \to Y$ can be factored uniquely as an epimorphism followed by a subspace inclusion:

$$\begin{array}{ccc} X & \xrightarrow{f} & Y \\ & \searrow\!\!\searrow \quad \nearrow & \\ & f(X) & \end{array}$$

The lattice $\mathcal{O}(f(X))$ is the set of all closed elements for the local operator $f_* f^-\colon \mathcal{O}(Y) \to \mathcal{O}(Y)$. We say $f(X)$ is the image of f. It is the smallest subspace of Y through which f factors. More generally, if $X_\ell \hookrightarrow X$ is a subspace of X, its image $f(X_\ell)$ is the subspace of Y determined by the local operator $f_* \ell f^-\colon \mathcal{O}(Y) \to \mathcal{O}(Y)$.

2. Points and discrete spaces

The lattice of open parts of the terminal object $1 \in Sp$ is the initial locale:

$$\mathcal{O}(1) = P1.$$

If $X \in Sp$, a continuous map $1 \to X$ is called a point of X. We write $|X|$ for $Sp(1,X)$, the set of points of X. We have

$$|X| = Loc(\mathcal{O}(X),\ P1),$$

but a morphism $t\colon \mathcal{O}(X) \to P1$ of locales is nothing but the characteristic map of a strongly prime filter $F \subseteq \mathcal{O}(X)$, i.e. $t = \chi_F$, where $F = t^{-1}(1)$. F satisfies:

i) $u \geq v \in F \Rightarrow u \in F$

ii) $u,v \in F \Rightarrow u \wedge v \in F$

iii) $1 \in F$

iv) $\bigvee_{i \in I} u_i \in F \Rightarrow \exists i \in I \ \ u_i \in F$.

For $I \in S$, consider $\coprod_I 1$ - the coproduct of I copies of 1. The continuous mappings $\coprod_I 1 \to X$ are in an one to one correspondence with the mappings $I \to Sp(1,X) = |X|$. We say $\coprod_I 1$ is the <u>discrete</u> space on I and we write I_{dis} for it. We have:

$$\mathcal{O}(I_{dis}) = \mathcal{O}(\coprod_I 1) = \mathcal{O}(1)^I = (P1)^I = P(I)$$

and in

$$Sp \underset{|\,|}{\overset{(\ )_{dis}}{\rightleftarrows}} S$$

we have $(\ )_{dis} \dashv |\ |$

<u>Proposition 1</u>. $(\ )_{dis}$ is full and faithful.

<u>Proof</u>: $(\ )_{dis}$ is full and faithful iff the unit of the adjunction $(\ )_{dis} \dashv |\ |$ is an isomorphism:

$$I \xrightarrow[\simeq]{} |I_{dis}| = Loc(PI, P1).$$

So we want to show that any strong prime filter F on PI is principal. But $I \in F$ and $I = \bigcup_{i \in I} \{i\}$, so we can find $i \in I$ such that $\{i\} \in F$. Suppose $J \in F$. Then $J \cap \{i\} \in F$ and the above shows that $\exists j \in J \cap \{i\}$ such that $\{j\} \in F$. Thus, $\{i\} \subseteq J$, and F is the principal filter generated by $i \in I$.

3. <u>The Sierpinski space</u>

Let $\$ \in Sp$ be defined by

$$\mathcal{O}(\$) = L(1),$$

the free locale on 1 generator. Specifically, we have

$$\mathcal{O}(\$) = \mathcal{D}(K(1)^0) = \mathcal{D}(2).$$

That is, $\mathcal{O}(\$)$ is the lattice of downward closed subsets of the partial order

$$2 = \{\bullet \longrightarrow \bullet\}.$$

Clearly, $Sp(X,\$) = Loc(L(1), \mathcal{O}(X)) = \mathcal{O}(X)$. This can be formulated as follows: let $s \in \mathcal{O}(\$)$ be the free generator; for any $u \in \mathcal{O}(X)$ there is exactly one continuous map $f: X \to \$$ such that $f^{-}(s) = u$. $\$$ is called the Sierpinski space.

The free locale $L(I)$ on I is the coproduct of I copies of $L(1)$. Thus $\mathcal{O}(\$^I) = L(I)$. We conclude that any space X can be expressed as the equalizer of a pair $\$^J \rightrightarrows \$^I$ of continuous maps.

4. Pullbacks and projective limits

The locale of open parts of the product $X\times Y$ is the tensor product $\mathcal{O}(X)\otimes\mathcal{O}(Y)$. More generally, if $X \to Z$ and $Y \to Z$ is a pair of continuous maps with common codomain, then

$$\mathcal{O}(X \underset{Z}{\times} Y) = \mathcal{O}(X) \underset{\mathcal{O}(Z)}{\otimes} \mathcal{O}(Y).$$

Proposition 1. Let $(X_i \to Z)_{i\in I}$ be a family of continuous maps, and $Y \to Z$ a continuous map from Y to Z. Then the canonical map

$$\coprod_{i\in I} (Y \underset{Z}{\times} X_i) \xrightarrow{\ \simeq\ } Y \underset{Z}{\times} (\coprod_{i\in I} X_i)$$

is an isomorphism.

Proof: We have to show that the canonical mapping

$$\mathcal{O}(Y) \underset{\mathcal{O}(Z)}{\otimes} \prod_{i\in I} \mathcal{O}(X_i) \to \prod_{i\in I} (Y) \underset{\mathcal{O}(Z)}{\otimes} \mathcal{O}(X_i)$$

is an isomorphism, but this is a consequence of Proposition 2, Chapter I, §5.

We have the remarkable fact:

Proposition 2. Let $(X_i, f_{ij})_{i\in I}$ be a projective system of spaces indexed by a filtered partially ordered set I. Suppose that for any $i \leq j$, the map $f_{ij}: X_i \to X_j$ is an epimorphism. Then, for any $j \in I$, the canonical projection $f_j: \varprojlim_{i\in I} X_i \to X_j$ is an epimorphism.

Proof: This is the dual of Proposition 1 Chapter III §2.

Note the striking difference here with the theory of classical spaces, even when AC is satisfied in S, where an instructive counterexample is given by the poset of finite partial injections from a large set to a smaller infinite one.

5. <u>The splitting space</u>

If X is a space, write $h\colon X' \to X$ for the dual of the solution $\mathcal{O}(X) \to \mathcal{O}(X)'$ to the problem of adding complements to every element of $\mathcal{O}(X)$. We say X' is the <u>splitting space</u> of X. The map h is epimorphic and monomorphic.

<u>Proposition 1</u>. For any subspace $S \hookrightarrow X$, the pullback $h^{-1}(S) \hookrightarrow X'$ is closed, isomorphic to S', and $h(S') = S$. Moreover, any closed subspace of X' is of the form $h^{-1}(S)$ for some subspace $S \hookrightarrow X$.

<u>Proof</u>: $\mathcal{O}(S)$ is the quotient of $\mathcal{O}(X)$ by the congruence relation generated by a family $(u_i, v_i)_{i\in I}$ of pairs of elements of $\mathcal{O}(X)$. $\mathcal{O}(h^{-1}(S))$ is the quotient of $\mathcal{O}(X)'$ by the congruence relation generated by the family of pairs $(h^{-}u_i, h^{-}v_i)_{i\in I}$. But the conditions $h^{-}u_i \equiv h^{-}v_i$ are equivalent to the conditions

$$((h^{-}u_i) \wedge (h^{-}v_i)') \vee ((h^{-}u_i)' \wedge (h^{-}v_i)) \equiv 0 \quad \forall i \in I,$$

or to the single condition

$$\bigvee_{i\in I} (h^{-}u_i) \,\Delta\, (h^{-}v_i) \equiv 0,$$

where Δ is the symmetric difference above. Thus $h^{-1}(S)$ is closed. Any element of $\mathcal{O}(S)$ becomes complemented in $\mathcal{O}(h^{-1}(S))$, since $\mathcal{O}(X) \to \mathcal{O}(S)$ is surjective, and every element in $\mathcal{O}(X)$ becomes complemented in $\mathcal{O}(X)'$. This shows $h^{-1}(S) = S'$ - just check the universal property. Also, in the pullback square

$$\begin{array}{ccc} h^{-1}(S) & \longrightarrow & S \\ \big\downarrow & & \big\downarrow \\ X' & \xrightarrow{\;h\;} & X \end{array},$$

the top line is an epimorphism, since $h^{-1}(S) = S'$. But then $h(h^{-1}(S)) = S$. Finally, a closed subspace of X' is described by a condition $v \equiv 0$ for some $v \in \mathcal{O}(X)'$. But v can be expressed as a supremum of elements of the form $h^{-}(u_i) \wedge h^{-}(v_i)'$ for $u_i, v_i \in \mathcal{O}(X)$ by Proposition 2 Chapter III §5:

$$v = \bigvee_{i\in I} h^{-}(u_i) \wedge h^{-}(v_i)'.$$

Going backwards, this means that the closed subspace CV is equal to $h^{-1}S$, where $\mathcal{O}(S)$ is the quotient of $\mathcal{O}(X)$ by the congruence relation generated by the set of pairs $(u_i, u_i \wedge v_i)_{i\in I}$.

Thus, we have

<u>Theorem 1</u>. The lattice of all subspaces of X is isomorphic to the lattice of closed subspaces of X', i.e. to $\mathcal{O}(X)'^{0}$.

In particular, we obtain the result of Dowker and Papert [10], and Isbell [11]: the lattice of local operators on a locale $\mathcal{O}(X)$ is itself a (0-dimensional in the obvious sense) locale - namely $\mathcal{O}(X)'$.

CHAPTER V - OPEN MAPS OF SPACES

In this chapter we introduce the central concept of an <u>open mapping</u> between spaces. We develop its basic properties, and show that it behaves in a way closely parallel to the concept of open map between topological spaces. Finally, we use openness to characterize discrete spaces and define local homeomorphisms (etale spaces).

1. <u>Open maps - definition</u>

<u>Proposition 1</u>. Let $f\colon X \to Y$ be a continuous map of spaces, and suppose $f^{-}\colon O(Y) \to O(X)$ has a left adjoint $\exists_f\colon O(X) \to O(Y)$. Then the following three conditions are equivalent.

1) $\forall y \in O(Y)$, $\forall x \in O(X)$,

$$\exists_f(f^{-}(y) \wedge x) = y \wedge \exists_f(x).$$

2) $\forall y,z \in O(Y)$,

$$f^{-}(y \to z) = f^{-}(y) \to f^{-}(z)$$

3) $\forall y \in O(Y)$, $\forall x \in O(X)$

$$f_*(x \to f^{-}(y)) = \exists_f(x) \to y.$$

<u>Proof</u>: The first condition means that $\forall y \in O(Y)$, the square

$$\begin{array}{ccc} O(X) & \xrightarrow{\exists_f} & O(Y) \\ {\scriptstyle f^{-}(y)\wedge(\ )}\downarrow & & \downarrow{\scriptstyle y\wedge(\ )} \\ O(X) & \xrightarrow{\exists_f} & O(Y) \end{array}$$

commutes. But this square commutes iff the square of right adjoints

$$\begin{array}{ccc} O(X) & \xleftarrow{f^{-}} & O(Y) \\ {\scriptstyle f^{-}(y)\to(\ )}\uparrow & & \uparrow{\scriptstyle y\to(\ )} \\ O(X) & \xleftarrow{f^{-}} & O(Y) \end{array}$$

commutes, which is condition 2).

Similarly, condition 1) can be expressed by saying that $\forall x \in O(X)$, the square

$$\begin{array}{ccc} O(X) & \xleftarrow{f^-} & O(Y) \\ {\scriptstyle (\ )\wedge x}\downarrow & & \downarrow{\scriptstyle \exists_f(x)\wedge(\ )} \\ O(X) & \xrightarrow{\exists_f} & O(Y) \end{array}$$

commutes, and again, this is so iff the corresponding square of right adjoints

$$\begin{array}{ccc} O(X) & \xrightarrow{f_*} & O(Y) \\ {\scriptstyle x\to(\ )}\uparrow & & \uparrow{\scriptstyle \exists_f(x)\to(\ )} \\ O(X) & \xleftarrow[f^-]{} & O(Y) \end{array}$$

commutes, which is condition 3).

Definition. An open mapping $f: X \to Y$ is a continuous map such that $f^-: O(Y) \to O(X)$ has a left adjoint $\exists_f: O(X) \to O(Y)$ satisfying any one of the equivalent conditions of Proposition 1.

For classical topological spaces, $\exists_f$ is, of course, direct image.

2. Open subspaces

Proposition 1. A subspace $i: S \hookrightarrow X$ is open iff i is an open mapping.

Proof: If $S \hookrightarrow X$ is an open subspace, then $O(S) = O(U)$ for some $u \in O(x)$, in which case the inclusion $O(U) \hookrightarrow O(X)$ is left adjoint to the projection $(\ ) \wedge u: O(X) \to O(U)$ and it clearly satisfies condition 1) of §1 Proposition 1. Conversely, suppose $i: S \hookrightarrow X$ is an open mapping. $i^-: O(X) \to O(S)$ being surjective, $O(S)$ is isomorphic to the lattice of coclosed elements for the coclosure operator $\exists_i i^-: O(X) \to O(X)$. But

$$\exists_i i^-(x) = \exists_i(i^-(x) \wedge 1) = x \wedge \exists_i(1).$$

Thus $\exists_i i^-(x) = x$ iff $x \leq \exists_i(1)$ so $O(S) = O(\exists_i(1))$.

Corollary. Let $(X_i \xrightarrow{f_i} Y_i)_{i\in I}$ be a family of continuous maps. Then for any $i \in I$, the canonical map $\mu_i: X_i \to \coprod_{i\in I} X_i$ is the inclusion of an open subspace, and

$$\begin{array}{ccc} X_i & \xrightarrow{f_i} & Y_i \\ {\scriptstyle \mu_i}\downarrow & \coprod_{i\in I} f_i & \downarrow{\scriptstyle \mu_i} \\ \coprod_{i\in I} X_i & \longrightarrow & \coprod_{i\in I} Y_i \end{array}$$

is a pullback.

<u>Proof</u>: The projection $\mu_i^-: \prod_{i\in I} \mathcal{O}(X_i) \to \mathcal{O}(X_i)$ preserves the binary $\to$ and has a left adjoint $\Sigma_{\mu_i}: \mathcal{O}(X_i) \to \prod_{i\in I} \mathcal{O}(X_i)$ (cf. Chapter I §2 Proposition 1). Thus, μ_i is an open mapping. The same result shows that $\mu_i^-\Sigma_{\mu_i} = 1_{\mathcal{O}(X_i)}$, so μ_i is an inclusion. Finally, the diagram

$$\begin{array}{ccc} \mathcal{O}(X_i) & \xleftarrow{f_i^-} & \mathcal{O}(Y_i) \\ {\scriptstyle \Sigma_{\mu_i}}\downarrow & & \downarrow{\scriptstyle \Sigma_{\mu_i}} \\ \prod_{i\in I} \mathcal{O}(X_i) & \xleftarrow[f^- = \prod_{i\in I} f_i^-]{} & \prod \mathcal{O}(Y_i) \end{array}$$

commutes, since the Σ_{μ_i} are the canonical injections in a coproduct of sup-lattices. Hence, $f^-(\Sigma_{\mu_i}(1)) = \Sigma_{\mu_i}(1)$, from which the final statement follows: any diagram of the form

$$\begin{array}{ccc} \mathcal{O}(W) & \xrightarrow{q^-} & \mathcal{O}(Z) \\ \downarrow & & \downarrow \\ \mathcal{O}(U) & \longrightarrow & \mathcal{O}(V) \end{array}$$

where $u \in \mathcal{O}(W)$, $v \in \mathcal{O}(Z)$ and $q^-(u) = v$, is a pushout of locales.

3. <u>Conditions for openness</u>

<u>Proposition 1</u>. For any space X, the map $p: X \to 1$ is open iff $p^-: P(1) \to \mathcal{O}(X)$ has a left adjoint.

<u>Proof</u>: Suppose p^- has a left adjoint $\exists: \mathcal{O}(X) \to P(1)$. We have to prove that for any $y \in P(1)$ and $x \in \mathcal{O}(X)$,

$$y \wedge \exists(x) = \exists(p^-(y) \wedge x).$$

The usual argument (Chapter III §4 Proposition 3) shows that it is enough to prove that the right hand side is 1 when the left hand side is 1. But in that case we have: $y = 1$ and $\exists(x) = 1$, so the right hand side is

$$\exists(p^-(y) \wedge x) = \exists(1 \wedge x) = \exists(x) = 1.$$

We call X an <u>open space</u> if $X \to 1$ is open.

Proposition 1, together with the use of covering systems (Chapter III §4), gives many examples of open spaces. In fact, let P be a partially ordered set equipped with a covering system $\mathrm{Cov}(x) \subseteq P{\downarrow}(x)$ for each $x \in P$. Let Q denote the locale quotient of $\mathcal{D}(P)$ by the congruence relation generated by the set of pairs $(\hat{R}, {\downarrow}(x))$ for $x \in P$ and $R \in \mathrm{Cov}(x)$. ($\hat{R}$ denotes the downward closed subset of P generated by R as in Chapter III §4.) Then we have:

<u>Proposition 2</u>. If for each $x \in P$ and each $R \in \mathrm{Cov}(x), \exists y \in R$, then Q represents an open space.

<u>Proof</u>: Consider the diagram

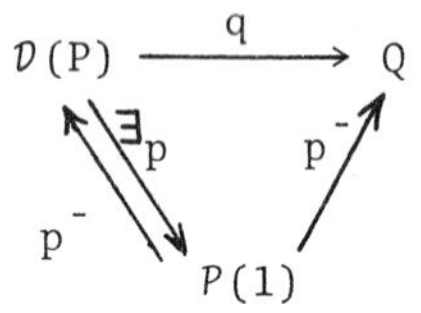

where q is the quotient map, each p^- is canonical, and $\exists_p$ is the usual existentiation of $\mathcal{P}(P)$ restricted to $\mathcal{D}(P)$. Clearly, $\exists_p \dashv p^-$, i.e. $\mathcal{D}(P)$ represents an open space. An easy calculation shows that if $\exists_p$ passes to a morphism $\exists\colon Q \to \mathcal{P}(1)$, then $\exists \dashv p^-$. But as we saw at the end of Chapter III §4, the locale quotient of $\mathcal{D}(P)$ is the same as the <u>sup-lattice quotient</u> of $\mathcal{D}(P)$ by the sup-lattice congruence generated by the set of pairs $(\hat{R}, {\downarrow}(x))$ for $x \in P$ and $R \in \mathrm{Cov}(x)$. Thus, $\exists_p$ passes to a sup-lattice morphism $\exists$ iff $\exists_p\hat{R} = \exists_p{\downarrow}(x)$ for each $x \in P$ and $R \in \mathrm{Cov}(x)$. However, as a morphism of sup-lattices, $\exists_p$ is determined by the condition $\exists_p{\downarrow}(x) = 1$ for each $x \in P$. So $\exists_p$ passes to Q iff $\exists_p\hat{R} = 1$ for each $R \in \mathrm{Cov}(x)$. Since $\exists_p\hat{R} = 1$ iff $\exists_p R = 1$, we are done. Note that the proof supplies also a partial converse. Namely, we can say that if Q represents an open space and if $\exists \circ q = \exists_p$, then for each $x \in P$ and each $R \in \mathrm{Cov}(x), \exists y \in R$.

As an important example, consider the problem of <u>enumerating</u> an arbitrary $X \in \mathcal{S}$. Let $\mathbb{N}$ denote the natural number object. We construct first the lattice of open parts of the space of functions from $\mathbb{N}$ to X. Let P be the poset of non-empty finite sequences $[x_0,\ldots,x_n]$ where $x_i \in X$, and $[x_0,\ldots,x_n, x_{n+1},\ldots,x_m] \leq [x_0,\ldots,x_n]$. The reader should think of $[x_0,\ldots,x_n]$ as the set of functions $f\colon \mathbb{N} \to X$ such that $f(i) = x_i$ $0 \leq i \leq n$. We say that $\{[x_0,\ldots,x_n,x_{n+1}] \mid x_{n+1} \in X\}$ covers $[x_0,\ldots,x_n]$. This is a covering system, and using the notation of Proposition 2, it is clear that Q represents an open space - the x_i's can always be repeated in the covers. An easy calculation shows that a point of Q is nothing but a function from $\mathbb{N}$ to X, so sheaves(Q) classifies these functions. Now, to form the space of epimorphisms from $\mathbb{N}$ to X, we simply extend the covers by specifying that, in addition, $\forall x \in X\{[x_0,\ldots,x_n,x_{n+1},\ldots,x_m] \mid x_{n+i} \in X \quad 1 \leq i \leq m,\ m \geq 0$ and $x \in \{x_0,\ldots,x_n,x_{n+1},\ldots,x_m\}\}$ covers $[x_0,\ldots,x_n]$. Again, it is immediate that this is a covering system, and a point of the new Q is now an epimorphism from $\mathbb{N}$ to X. This Q still represents an open space, since, for example, $[x_0,\ldots,x_n,x]$ is in the new cover of $[x_0,\ldots,x_n]$ corresponding to x.

Returning to characterizations of openness, we have

Proposition 3. A continuous map $f\colon X \to Y$ is open iff the image of any open subspace $U \hookrightarrow X$ is open.

Proof: An open subspace $U \hookrightarrow X$ is described by the local operator $u \to (\ )$ on $\mathcal{O}(X)$. The image $f(U) \hookrightarrow Y$ is described by the local operator $f_*(u \to f^-(\ ))$ (Chapter IV §1). If f is open, then we have $f_*(u \to f^-(\ )) = \exists_f(u) \to (\ )$, so $f(U) \hookrightarrow Y$ is the open subspace defined by $\exists_f(u) \in \mathcal{O}(Y)$. Conversely, if $f(U)$ is open, we can write

$$f_*(u \to f^-(y)) = f(u) \to y,$$

where $f(u) \in \mathcal{O}(Y)$. Then

$$\dfrac{\dfrac{u \leq f^-(y)}{1 = (u \to f^-(y))}}{\dfrac{1 = f_*(u \to f^-(y))}{\dfrac{1 = (f(u) \to y)}{f(u) \leq y}}}$$

showing that f^- has a left adjoint satisfying condition 3) of §1 Proposition 1.

The interior $\mathring{S}$ of a subspace $S \hookrightarrow X$ is the largest open subspace contained in S.

Proposition 4. A continuous map $f\colon X \to Y$ is open iff for any subspace $S \hookrightarrow Y$, we have

$$f^{-1}(\mathring{S}) = \overset{\circ}{\widehat{f^{-1}(S)}}\ .$$

Proof: Suppose f is open. For any open subspace $U \hookrightarrow X$ we have

$$\begin{array}{c} U \leq f^{-1}(\mathring{S}) \\ \hline f(U) \leq \mathring{S} \\ \hline f(U) \leq S \\ \hline U \leq f^{-1}(S) \\ \hline U \leq \overset{\circ}{\widehat{f^{-1}(S)}} \end{array} ,$$

showing that $f^{-1}(\mathring{S}) = \overset{\circ}{\widehat{f^{-1}(S)}}$. Conversely, suppose this is true, and let $U \hookrightarrow X$ be open. We have

$$\begin{array}{c} f(U) \leq S \\ \hline U \leq f^{-1}(S) \\ \hline U \leq \overset{\circ}{\widehat{f^{-1}(S)}} \\ \hline U \leq f^{-1}(\mathring{S}) \\ \hline f(U) \leq \mathring{S} \end{array} ,$$

showing that $f(U)$ is open.

4. Open Surjections, pullbacks

Let $f\colon X \to Y$ be an open mapping. For any $u \in \mathcal{O}(Y)$ we have $\exists_f(f^{-}(u)) = \exists_f(f^{-}(u) \wedge 1) = u \wedge \exists_f(1) = u \wedge f(X)$. When $\exists_f(1) = 1$, f is an epimorphism, and we call f an open surjection.

We say T is an open, surjective space when $T \to 1$ is an open surjection. Note that the Q of §3 Proposition 2 represents an open surjective space iff it represents an open space, and $\exists x \in P$. Thus, in the enumeration problem for an $X \in \mathcal{S}$, the space of epimorphisms from $\mathbb{N}$ to X is an open surjective space iff $\exists x \in X$. Hence, even though this space may have no points over $\mathcal{S}$, it is still highly non-trivial when X has global support. The possibility of consistent enumeration will play a fundamental role later on.

Proposition 1. Let $f\colon X \to Z$ and $g\colon Y \to Z$ be continuous mappings. If f is open, then the pullback f' of f along g is open.

$$\begin{array}{ccc} Y \times_Z X & \xrightarrow{\;h\;} & X \\ {\scriptstyle f'}\downarrow & & \downarrow{\scriptstyle f} \\ Y & \xrightarrow{\;g\;} & Z \end{array}$$

Moreover, for any $x \in \mathcal{O}(X)$, we have

$$g^{-}\exists_f(x) = \exists_{f'}h^{-}(x).$$

In particular, if f is surjective, so is f'.

Proof: $\mathcal{O}(Y\times_Z X) = \mathcal{O}(Y) \otimes_{\mathcal{O}(Z)} \mathcal{O}(X)$, where $\mathcal{O}(Y)$ and $\mathcal{O}(X)$ are equipped with the usual $\mathcal{O}(Z)$ structures. The identity $\exists_f(f^{-}(z) \wedge x) = z \wedge \exists_f(x)$ means exactly that $\exists_f\colon \mathcal{O}(X) \to \mathcal{O}(Z)$ is a morphism of $\mathcal{O}(Z)$-modules. Thus, we get a commutative diagram

$$\begin{array}{ccc} \mathcal{O}(Y) \otimes_{\mathcal{O}(Z)} \mathcal{O}(X) & \xleftarrow{\;h^{-}\;} & \mathcal{O}(X) \\ {\scriptstyle 1\otimes\exists_f}\big\downarrow & & \big\downarrow{\scriptstyle \exists_f} \\ \mathcal{O}(Y) & \xleftarrow{\;g^{-}\;} & \mathcal{O}(Z) \end{array} .$$

Moreover, $1\otimes\exists_f$ is a left adjoint to $1\otimes f^{-}$ since the adjunctions $1 \leq f^{-}\exists_f$ and $\exists_f f^{-} \leq 1$ are preserved under $\otimes$-product. The condition $\exists_{f'}(f'^{-}(y) \wedge u) = y \wedge \exists_{f'}(u)$ for $y \in \mathcal{O}(Y)$ and $u \in \mathcal{O}(Y\times_Z X)$ means that $1\otimes\exists_f$ is $\mathcal{O}(Y)$ linear, which is true because $\exists_f$ is $\mathcal{O}(Z)$-linear.

Proposition 2. Let $p\colon X \twoheadrightarrow Z$ be an open surjection. Then p is the coequalizer of the pair $X\times_Z X \underset{p_2}{\overset{p_1}{\rightrightarrows}} X$.

Proof: We must show that

$$\mathcal{O}(Z) \rightarrowtail^{\;p^{-}\;} \mathcal{O}(X) \underset{p_2^{-}}{\overset{p_1^{-}}{\rightrightarrows}} \mathcal{O}(X\times_Z X)$$

is an equalizer. So let $u \in \mathcal{O}(X)$ be such that $p_1^{-}(u) = p_2^{-}(u)$. We have

$$\begin{aligned} p^{-}(\exists_p(u)) &= \exists_{p_2}(p_1^{-}(u)) \\ &= \exists_{p_2}(p_2^{-}(u)) \\ &= u , \end{aligned}$$

since p_2 is an open surjection, and this proves the result.

5. A characterization of discrete spaces

Here we prove

Theorem 1. A space X is discrete iff the canonical projection $p: X \to 1$ and the diagonal $\Delta: X \to X \times X$ are open.

First we need two lemmas.

Lemma 1. A space Y is isomorphic to 1 iff the canonical projection $p: Y \to 1$ and the diagonal $\Delta: Y \to Y \times Y$ are open surjections.

Proof: If $p: Y \to 1$ is an open surjection, it is the coequalizer of the two projections $p_1, p_2: Y \times Y \to Y$ by §4 Proposition 2. By the same result, if Δ is an open surjection, $p_1 = p_2$, p is monic, and the lemma follows.

Lemma 2. Suppose $p: X \to 1$ is open, and $u \in \mathcal{O}(X)$ satisfies $u = \bigvee_{i \in I} u_i$. Let $J = \{i \mid \exists u_i = 1\}$. Then $u = \bigvee_{i \in J} u_i$.

Proof: We have to show: $\forall i \in I,\ u_i \leq \bigvee_{j \in J} u_j$. But $u_i \leq p^{-}\exists u_i$, so it is enough to show

$$\frac{\dfrac{u_i \wedge p^{-}\exists u_i \leq \bigvee_{j \in J} u_j}{p^{-}\exists u_i \leq u_i \to \bigvee_{j \in J} u_j}}{\exists u_i \leq p_*(u_i \to \bigvee_{j \in J} u_j)}$$

By the usual argument, it is enough to show that if the left hand side of the bottom line equals 1, then the right hand side equals 1, but this is obvious.

Proof of Theorem 1: An atom of X is an open subspace $a \hookrightarrow X$ such that $a \times a \subseteq \Delta$ and $\exists a = 1$. Let A be the set of atoms. Each atom a defines a point of X, since $a \simeq 1$ by Lemma 1. Define

$$\phi: \mathcal{O}(X) \to P(A)$$

by $\phi(u) = \{a \in A \mid a \leq u\}$. Clearly, $\phi(u \wedge v) = \phi(u) \cap \phi(v)$, and $\phi(1) = A$. Also, however, $\phi(\bigvee_{i \in I} u_i) = \bigcup_{i \in I} \phi(u_i)$, because, since $a \simeq 1$, $a \leq \bigvee_{i \in I} u_i \Longrightarrow \exists i \in I\ \ a \leq u_i$. So ϕ is a morphism of locales. Furthermore, ϕ has a left adjoint $\sigma: P(A) \to \mathcal{O}(X)$ defined by

$$\sigma(I) = \bigvee\{a \mid a \in I\}.$$

Clearly,

$$\frac{\sigma(I) \leq u}{I \leq \phi(u)} .$$

We have $\phi\sigma(I) = I$, for by adjointness $I \subseteq \phi\sigma(I)$, and if $a' \in \phi\sigma(I)$, then $a' \leq \bigvee\{a \mid a \in I\} \Longrightarrow \exists a \in I \;\; a' \leq a \Longrightarrow a' = a$, since a' and a are both isomorphic to 1.

On the other hand, we claim also, $\sigma\phi(u) = u$, for $U \hookrightarrow X \overset{\Delta}{\hookrightarrow} X\times X$, and Δ being open, as an open subspace of $X\times X$ we have

$$\Delta(U) = \bigvee\{v\times w \mid v,w \in \mathcal{O}(X) \;\&\; v\times w \subseteq \Delta(U)\}.$$

But $v\times w \subseteq \Delta(U) \Longrightarrow v\times w = w\times v$, so

$$\Delta(U) = \bigvee\{v\times v \mid v \in \mathcal{O}(X) \;\&\; v\times v \subseteq \Delta(U)\}.$$

Thus, $u = \bigvee\{v \in \mathcal{O}(X) \mid v\times v \subseteq \Delta(u)\}$, and we may apply Lemma 2 to get the result.

Definition. An etale space over X is a continuous map $p\colon E \to X$, which is discrete as a space over X. That is, p is open, and $\Delta\colon E \to E\times_X E$ is open.

In the next chapter, we will see that the topos of sheaves on X is equivalent to the category of etale spaces and continuous maps over X.

CHAPTER VI - CHANGE OF BASE

One of the main advantages of mathematics done over an arbitrary base topos is the possibility of <u>changing the base</u> along a morphism. Thus, we are led, in analogy with the situation induced by a continuous map of spaces, to consider topoi E and S connected by a geometric morphism $p: E \to S$ consisting of a pair of functors

$$E \underset{p_*}{\overset{p^*}{\rightleftarrows}} S$$

with p^* left exact and left adjoint to p_*. We suppose E can be expressed as the category of S-valued sheaves on some site in S. In the present chapter we examine the relations between sup-lattices and locales in E and S respectively.

Our first task will be to lift the adjoint pair $p^* \dashv p_*$ to sup-lattices and locales and then to establish various properties of these lifted pairs. Next, we describe sup-lattices and locales in E as certain structures in S. For example, if E is the category of sheaves on a locale A in S, then the category $s\ell(E)$ of sup-lattices in E is equivalent to the category Mod(A) of A-modules in S. Similarly, Loc(E) - the category of locales in E - is equivalent to the category of locale extensions of A in S.

Using these descriptions, we study relations between the lifted adjoints in the case of a pullback square of topoi. The resulting formulas provide an important technique for investigating the stability of various kinds of geometric morphisms under change of base.

1. <u>Change of base for sup-lattices and locales</u>

We begin with a few preliminary remarks. Namely, let $p: E \to S$ be a topos defined over S. If $M \in s\ell(E)$ and $I \in E$, then $M^I \in s\ell(E)$. Furthermore, if $\alpha: I \to J$, then $a^* = M^\alpha: M^J \to M^I$ has a left adjoint $\Sigma_\alpha: M^I \to M^J$ defined by

$$\Sigma_\alpha(f)(j) = \bigvee_{\alpha(i)=j} f(i).$$

Moreover, for any pullback square

$$\begin{array}{ccc} I' & \xrightarrow{\beta'} & I \\ {\scriptstyle\alpha'}\downarrow & & \downarrow{\scriptstyle\alpha} \\ J' & \xrightarrow{\beta} & J \end{array}$$

in E we have

$$\Sigma_{\alpha'}\beta'^* = \beta^*\Sigma_\alpha,$$

as is immediate from the above formula. Similarly if $A \in \mathrm{Loc}(E)$ and $I \in E$, then $A^I \in \mathrm{Loc}(E)$. In addition, if $\alpha: I \to J$, the adjoint pair $\Sigma_\alpha \dashv \alpha^*$ satisfies Frobenius reciprocity i.e. for $f \in A^I$ and $g \in A^J$,

$$\Sigma_\alpha(\alpha^*(g) \wedge f) = g \wedge \Sigma_\alpha(f),$$

which also follows by direct calculation from the formula for Σ_α.

Proposition 1. The direct image $p_*(M)$ of any sup-lattice M in E is a sup-lattice in S. The direct image $p_*(f)$ of a morphism $f: M \to N$ of $s\ell(E)$ is a morphism of $s\ell(S)$. Moreover, the induced functor $p_*: s\ell(E) \to s\ell(S)$ has a left adjoint $p^\#: s\ell(S) \to s\ell(E)$.

Proof: We can calculate the supremum of a family $I \to p_*(M)$ as the supremum of the corresponding family $p^*(I) \to M$. The preservation of suprema by $p_*(f)$ is proved by the same method. It remains to show the existence of $p^\#$.

Let $L \in s\ell(S)$, and express L as the coequalizer of a pair

$$PX \underset{s}{\overset{r}{\rightrightarrows}} PY \twoheadrightarrow L.$$

r and s are uniquely determined by their restriction to the free generators X of PX:

$$X \rightrightarrows PY,$$

and these, in turn, correspond to a pair of binary relations

$$R, S \subseteq X \times Y.$$

Applying p^*, we obtain relations

$$p^*R,\ p^*S \subseteq p^*X \times p^*Y,$$

hence two morphisms

$$\Omega^{p^*X} \rightrightarrows \Omega^{p^*Y}$$

of $s\ell(E)$. Their coequalizer is $p^\#(L)$.

Proposition 2. We have:

1) $\mathrm{Hom}(L,p_*(M)) \simeq p_*\mathrm{Hom}(p^{\#}(L),M)$

2) $p^{\#}(PX) \simeq \Omega^{p^*X}$

3) $p^{\#}(L\otimes M) \simeq p^{\#}(L)\otimes p^{\#}(M)$

4) $L\otimes p_*(M) \simeq p_*(p^{\#}(L)\otimes M)$.

Proof: The first identity expresses the (strong) adjointness $p^{\#} \dashv p_*$. 2) and 3) are immediate consequences of 1), and 4) follows from 1) using the formula $L\otimes M = \mathrm{Hom}(L,M^0)^0$.

The third identity of Proposition 2 implies that $p^{\#}$ preserves commutative monoids and locales, as does p_*. In fact, $p^{\#}$ is the left adjoint to $p_*\colon \mathrm{Loc}(E) \to \mathrm{Loc}(S)$. This will be important in the next chapter.

On the level of spaces, we adopt the convention

$$\mathcal{O}(p^*X) = p^{\#}\mathcal{O}(X)$$
$$\mathcal{O}(p_!X) = p_*\mathcal{O}(X),$$

thus $p_! \dashv p^*$ for spaces. Another way of writing the second identity of Proposition 2 is $p^*(I_{dis}) = (p^*I)_{dis}$ for $I \in S$, so this use of p^* for spaces is a consistent extension of its value on objects of S.

Proposition 3. p^* preserves open (surjective) spaces.

Proof: The adjointness $p^{\#} \dashv p_*$ is strong (identity 1) of Proposition 2), so $p^{\#}$ and p_* are both strong functors, and therefore $p^{\#}$ must preserve the left adjoint of Chapter V §3 Proposition 1.

Let A be a locale in S. We denote the category of S-valued sheaves on A by sh(A). If X is a space, a sheaf on X is a sheaf on $\mathcal{O}(X)$.

Proposition 4. If A is a locale of S, then

$$\begin{array}{ccc} \mathrm{sh}(p^{\#}A) & \xrightarrow{p'} & \mathrm{sh}(A) \\ \downarrow & & \downarrow \\ E & \xrightarrow{p} & S \end{array}$$

is a pullback square.

Proof: Diaconescu's Theorem says that morphisms of topoi

$$\begin{array}{ccc} F & \xrightarrow{f} & \mathrm{sh}(A) \\ & {\scriptstyle q}\searrow \quad \swarrow & \\ & S & \end{array}$$

are classified by left exact F valued functors on A which take covers to surjective families. These are exactly locale morphisms $A \to q_*\Omega$, so the Proposition follows immediately from the universal property of $p^\# A$. The morphism p' is induced by the adjunction $A \to p_* p^\# A$.

As an application of $p^\#$, and because we need the result in Chapter VII, we give here a new characterization of atomic topoi [5] over S.

Proposition 5. The following conditions on a morphism $p: A \to S$ are equivalent

1) p^* is logical

2) For each space X of S, $|p^*X| \simeq p^*|X|$.

3) For every $T \in A$, $p_!(T_{dis})$ is discrete.

Proof: If p^* is logical, then, since $|X| = \mathrm{Hom}(\mathcal{O}(X), \mathcal{P}(1))$, we have $p^*|X| = \mathrm{Hom}(p^*\mathcal{O}(X), \Omega)$. However, it follows from the construction of $p^\#$ given in Proposition 1 that if p^* is logical then $p^* = p^\#$. Thus $p^*|X| = |p^*X|$. On the other hand, if this holds for all spaces X, then it holds for $X = \$^I$, where $\$$ is the Sierpinski space (Chapter IV §3) and $I \in S$. But $|\$^I| = \mathrm{Hom}(L(I), \mathcal{P}(1)) = \mathcal{P}(1)^I$, and $|p^*(\$^I)| = \mathrm{Hom}(p^\# L(I), \Omega) = \mathrm{Hom}(L(p^*I), \Omega) = \Omega^{p^*I}$, so the identity $p^*|\$^I| = |p^*(\$^I)|$ simply says that p^* is logical. Thus 1) $\Longleftrightarrow$ 2) is established.

Suppose 3) holds, and X is a space in S. Since for each $T \in A$, $p_!(T_{dis})$ is discrete, we have

$$\frac{p_!(T_{dis}) \to X}{p_!(T_{dis}) \to |X|_{dis}}$$

But

$$\cfrac{\cfrac{\cfrac{p_!(T_{dis}) \to |X|_{dis}}{T_{dis} \to p^*(|X|_{dis})}}{T_{dis} \to (p^*|X|)_{dis}}}{T \to p^*|X|} ,$$

and

$$\cfrac{\cfrac{p_!(T_{dis}) \to X}{T_{dis} \to p^*X}}{T \to |p^*X|} ,$$

so $p^*|X| \simeq |p^*X|$. This argument is clearly reversible, so the Proposition is proved.

Although we will not prove it here, we remark that there is a characterization of locally connected topoi [6] similar to

Proposition 5 (2), where the arbitrary space X is replaced by an arbitrary product of discrete spaces.

2. Determination of $s\ell(S^{\underline{A}^{op}})$ and $Loc(S^{\underline{A}^{op}})$

Here we suppose $E = S^{\underline{A}^{op}}$ where $\underline{A}$ is a finitely complete category in S, and we give a complete description of sup-lattices and locales of E as functors $\underline{A}^{op} \to S$.

Proposition 1. A sup-lattice $M \in s\ell(S^{\underline{A}^{op}})$ is a functor $M: \underline{A}^{op} \to s\ell(S)$ satisfying the two conditions below:

1) For every $\alpha: A \to B$ in $\underline{A}$, the mapping $M(\alpha): M(B) \to M(A)$ has a left adjoint $\Sigma_\alpha: M(A) \to M(B)$.

2) For every pullback square

$$\begin{array}{ccc} A' & \xrightarrow{\beta'} & A \\ {\scriptstyle \alpha'}\downarrow & & \downarrow{\scriptstyle \alpha} \\ B' & \xrightarrow{\beta} & B \end{array} \quad ,$$

we have

$$\Sigma_{\alpha'} M(\beta') = M(\beta)\Sigma_\alpha .$$

Moreover, a morphism $\phi: M \to N$ of $s\ell(S^{\underline{A}^{op}})$ is a natural transformation such that for each $\alpha: A \to B$ of A,

$$\begin{array}{ccc} M(A) & \xrightarrow{\phi_A} & N(A) \\ {\scriptstyle \Sigma_\alpha}\downarrow & & \downarrow{\scriptstyle \Sigma_\alpha} \\ M(B) & \xrightarrow{\phi_B} & N(B) \end{array}$$

commutes.

Proof: If $Y: \underline{A} \to S^{\underline{A}^{op}}$ is the Yoneda embedding, we have the formulas:

$$p_*(M^{YA}) = M(A)$$

$$p_*(M^{Y\alpha}) = M(\alpha).$$

The preliminary remarks of §1 show that $M(A) \in s\ell(S)$, $M(\alpha)$ has a left adjoint - namely $p_*(\Sigma_{Y\alpha})$ - and condition 2) is satisfied.

Conversely, if $M: \underline{A}^{op} \to s\ell(S)$ is a functor satisfying 1) and 2), we will exhibit the supremum mapping $\sigma: \Omega^M \to M$. We have to give, for any $A \in \underline{A}$, a mapping $\sigma_A: \Omega^M(A) \to M(A)$. Replacing $\underline{A}$ by $\underline{A}/A$, we may suppose that $A = 1$. Therefore, it will be sufficient to calculate the supremum of a subfunctor $S \subseteq M$. Let

$$\overline{S}_A = \bigcup_{\alpha: B\to A} \Sigma_\alpha(S(B)),$$

where the union runs over all arrows $\alpha: B \to A$ of $\underline{A}$ with A fixed. Condition 2) implies $\overline{S} = (\overline{S}_A)_{A\in\underline{A}}$ is a subfunctor of M. For $A \in \underline{A}$, let $s_A = \sup \overline{S}_A$. $s_A \in M(A)$, we show it defines a global section: that is, for any $\alpha: A' \to A$ we show $M(\alpha)(s_A) = s_{A'}$. The inequality $M(\alpha)(s_A) \leq s_{A'}$ is a consequence of $M(\alpha)(\overline{S}_A) \leq \overline{S}_{A'}$, which holds since $M(\alpha)$ preserves suprema. The inequality $s_{A'} \leq M(\alpha)(s_A)$ is equivalent to $\Sigma_\alpha(s_{A'}) \leq s_A$. This is a consequence of $\Sigma_\alpha(\overline{S}_{A'}) \leq \overline{S}_A$, which holds since Σ_α preserves suprema. Finally, we show the global section $s = (s_A)_{A\in\underline{A}}$ is the supremum of $S \subseteq M$. Suppose that $t \in M(A)$ is an upper bound for S. This means that for every $\alpha: B \to A$ and every $x \in S(B)$, we have $x \leq M(\alpha)(t)$. Thus, $\Sigma_\alpha(x) \leq t$, so t is also an upper bound for $\overline{S}_A$. Therefore, $s_A \leq t$, and we are done.

Proposition 2. A locale $L \in \mathrm{Loc}(S^{\underline{A}^{op}})$ is a functor $L: \underline{A}^{op} \to \mathrm{Loc}(S)$ satisfying conditions 1) and 2) of Proposition 1, but, where, in addition, the left adjoints Σ_α of condition 1) satisfy Frobenius reciprocity:

$$\Sigma_\alpha(L(\alpha)(x) \wedge y) = x \wedge \Sigma_\alpha(y).$$

Proof: As in the proof of Proposition 1, the necessity of the conditions follows from the preliminary remarks of §1. Conversely, if $L: \underline{A}^{op} \to \mathrm{Loc}(S)$ satisfies the conditions, then the supremum in each $L(A)$ was calculated in the proof of Proposition 1, and the distributive law is a direct consequence of Frobenius reciprocity.

Suppose now that $\underline{A} = Z$ is an inf-semilattice. A Z-module is a sup-lattice M together with an operation

$$Z \times M \to M,$$

written $a \cdot x$ for $a \in Z$ and $x \in M$, such that

$$a \leq b \Longrightarrow a\cdot x \leq b\cdot x$$

$$a\cdot\left(\bigvee_{i\in I} x_i\right) = \bigvee_{i\in I} a\cdot x_i$$

$$a\cdot(b\cdot x) = (a\wedge b)\cdot x$$

$$1\cdot x = x \qquad .$$

Clearly, Z-module structures on M are in 1-1 correspondence with ordinary $\mathcal{D}(Z)$-module structures, where $\mathcal{D}(Z)$ is the free locale on the inf-semilattice Z (Chapter III §3).

Proposition 3. Let $M \in s\ell(S^{Z^{op}})$. The sup lattice $M(1) \in s\ell(S)$ is equipped with a canonical Z-module structure, and putting TM = M(1) defines an equivalence of categories

$$T: s\ell(S^{Z^{op}}) \to \mathrm{Mod}(Z)$$

Moreover, for any pair $M,N \in s\ell(S^{Z^{op}})$, we have a natural isomorphism

$$T(\mathrm{Hom}(M,N)) \simeq \mathrm{Hom}_Z(TM,TN).$$

Proof: Let $M \in s\ell(S^{Z^{op}})$, i.e. $M: Z^{op} \to s\ell(S)$ is a functor satisfying conditions 1) and 2) of Proposition 1. If $a \leq b$, denote the morphism $M(b) \to M(a)$ by ρ^b_a, and write Σ^b_a for its left adjoint. Then, the multiplication by a on TM = M(1) is given by the composite $\Sigma^1_a\rho^1_a$: $M(1) \to M(1)$. If $a,b \in Z$ consider the pullback square

$$\begin{array}{ccc} a \wedge b & \to & b \\ \downarrow & & \downarrow \\ a & \longrightarrow & 1 \end{array} .$$

By Proposition 1 condition 2), we have

$$\rho^1_a\Sigma^1_b = \Sigma^a_{a\wedge b}\rho^b_{a\wedge b}.$$

So, if $a \leq b$ and $x \in M(1)$,

$$\frac{\dfrac{a\cdot x \leq b\cdot x}{\Sigma^1_a\rho^1_a(x) \leq \Sigma^1_b\rho^1_b(x)}}{\rho^1_a(x) \leq \rho^1_a\Sigma^1_b\rho^1_b(x)}$$

$$= \Sigma^a_{a\wedge b}\rho^b_{a\wedge b}\rho^1_b(x)$$

$$= \Sigma^a_a\rho^b_a\rho^1_b(x)$$

$$= \rho^1_a(x),$$

and $a\cdot x \leq b\cdot x$ holds. For arbitrary a and b,

$$a\cdot(b\cdot x) = \Sigma^1_a\rho^1_a\Sigma^1_b\rho^1_b(x)$$

$$= \Sigma^1_a\Sigma^a_{a\wedge b}\rho^b_{a\wedge b}\rho^1_b(x)$$

$$= \Sigma^1_{a\wedge b}\rho^1_{a\wedge b}(x)$$

$$= (a \wedge b)\cdot x,$$

showing that TM is a Z-module.

On the other hand, let N be a Z-module, and define a functor $\tilde{N}: Z^{op} \to s\ell(S)$ by

$$\tilde{N}(a) = \{x \in N \mid a \cdot x = x\}.$$

For $a \leq b$, the sup-preserving mapping $\rho_a^b: \tilde{N}(b) \to \tilde{N}(a)$ is $x \longmapsto a \cdot x$, and its left adjoint $\Sigma_a^b: \tilde{N}(a) \to \tilde{N}(b)$ is the inclusion $\tilde{N}(a) \subseteq \tilde{N}(b)$. The conditions of Proposition 1 are clearly satisfied. Obviously,

$$T\tilde{N} = \tilde{N}(1) = N.$$

Moreover, considering the pullback square

$$\begin{array}{ccc} a & \to & a \\ \downarrow & & \downarrow \\ a & \to & 1 \end{array}$$

if $M: Z^{op} \to s\ell(S)$ is in $s\ell(S^{Z^{op}})$ it follows that

$$\rho_a^1 \Sigma_a^1 = \Sigma_a^a \rho_a^a = 1_{M(a)}.$$

But then

$$\begin{aligned} M(a) &\simeq \{x \in M(1) \mid \Sigma_a^1 \rho_a^1(x) = x\} \\ &= \widetilde{TM}(a), \end{aligned}$$

so $M \simeq \widetilde{TM}$. The rest of the proof is left to the reader.

3. Determination of $s\ell(sh(A))$ and $Loc(sh(A))$ for $A \in Loc(S)$

We begin with

Lemma 1. Let $E' \hookrightarrow E$ be a subtopos. A partially ordered object $M \in E'$ is a sup-lattice iff it is a sup-lattice in E.

Proof: Necessity is clear by Proposition 1 §1. Conversely, to calculate the supremum of an E'-subset of M, just calculate its supremum as an E-subset.

Proposition 1. Let A be a locale in S. Then there is a natural equivalence

$$s\ell(sh(A)) \simeq Mod(A).$$

Proof: Let N be an A-module. We have to show that the functor $\tilde{N}: A^{op} \to s\ell(S)$ of §2 is a sheaf. So let $a = \bigvee_{i \in I} a_i$ be a supremum in A, and let $(x_i)_{i\in I}$ be a family of elements $x_i \in \tilde{N}(a_i)$ such that for $i,j \in I$,

$$(a_i \wedge a_j)\cdot x_i = (a_i \wedge a_j)\cdot x_j$$

i.e.,

$$a_j\cdot x_i = a_i\cdot x_j.$$

Let $x = \bigvee_{i\in I} x_i$. Then

$$\begin{aligned} a\cdot x &= a\cdot \bigvee_{i\in I} x_i \\ &= \bigvee_{i\in I} a\cdot x_i \\ &= \bigvee_{i\in I} x_i = x, \end{aligned}$$

since $a_i \leq a$ yields $x_i = a_i\cdot x_i \leq a\cdot x_i \leq x_i$, so $a\cdot x_i = x_i$. Thus, $x \in \tilde{N}(a)$. Moreover,

$$\begin{aligned} a_i\cdot x &= a_i\cdot \bigvee_{j\in I} x_j \\ &= \bigvee_{j\in I} a_i\cdot x_j \\ &= \bigvee_{j\in I} a_j\cdot x_i \\ &= (\bigvee_{j\in I} a_j)\cdot x_i \\ &= a\cdot x_i = x_i, \end{aligned}$$

so x is a common extension of the elements of the family $(x_i)_{i\in I}$. To prove the uniqueness of the extension, suppose $y \in \tilde{N}(a)$ is such that $a_i\cdot y = x_i$ $\forall i \in I$. Then

$$\begin{aligned} y &= a\cdot y \\ &= (\bigvee_{i\in I} a_i)\cdot y \\ &= \bigvee_{i\in I} a_i\cdot y \\ &= \bigvee_{i\in I} x_i = x, \end{aligned}$$

and $\tilde{N}$ is a sheaf.

Conversely, let $M\colon A^{op} \to s\ell(S)$ be a sheaf. We have to show that the action $A\times M(1) \to M(1)$ is sup-preserving in the first variable - i.e. that

$$(\bigvee_{i \in I} a_i)\cdot x = \bigvee_{i \in I} a_i \cdot x.$$

Let $y \in M(1)$ be such that $\forall i \in I \quad a_i \cdot x \leq y$ - i.e.

$$\Sigma^1_{a_i} \rho^1_{a_i}(x) \leq y \qquad \text{or}$$

$$\rho^1_{a_i}(x) \leq \rho^1_{a_i}(y).$$

The partial order on M is a subsheaf of $M \times M$, so we can conclude that

$$\rho^1_a(x) \leq \rho^1_a(y)$$

where $a = \bigvee_{i \in I} a_i$. But then

$$\Sigma^1_a \rho^1_a(x) \leq y \qquad \text{or}$$

$$a \cdot x \leq y,$$

and we are done.

Proposition 2. The category $\mathrm{Loc}(\mathrm{sh}(A))$ is naturally equivalent to the category of locale extensions $A \to B$ of A in S.

Proof: The fact that the functor T of Proposition 3 §2 preserves Hom, means that in Proposition 1, it also preserves $\otimes$-product. Thus, T induces an equivalence between the category of commutative monoids of $s\ell(\mathrm{sh}(A))$ and the category of A-algebras in S. Locales correspond to locales, completing the proof.

Proposition 3. Let X be a space in S. Then the category of sheaves on X is equivalent to the category of etale spaces $E \to X$.

Proof: The category of sheaves on X is the category $\mathrm{sh}(\mathcal{O}(X))$, which is equivalent to the category of discrete spaces Y of $\mathrm{sh}(\mathcal{O}(X))$ which by Theorem 1 §5 Chapter V is the dual of the full subcategory of locales $\mathcal{O}(Y)$ in $\mathrm{sh}(\mathcal{O}(X))$ such that $\Omega \to \mathcal{O}(Y)$ and $\mathcal{O}(Y) \otimes \mathcal{O}(Y) \to \mathcal{O}(Y)$ have linear left adjoints. Under T these are taken to locale extensions

$$\mathcal{O}(X) \to \mathcal{O}(E)$$

such that $\mathcal{O}(X) \to \mathcal{O}(E)$ and $\mathcal{O}(E) \underset{\mathcal{O}(X)}{\otimes} \mathcal{O}(E) \to \mathcal{O}(E)$ have linear left adjoints. But this category is the dual of the category of etale spaces $E \to X$.

4. The Beck-Chevalley conditions

Theorem 1. For any pullback square of topoi

$$\begin{array}{ccc} E' & \xrightarrow{f'} & E \\ {\scriptstyle p'}\downarrow & & \downarrow{\scriptstyle p} \\ S' & \xrightarrow{f} & S \end{array},$$

in the square

$$\begin{array}{ccc} s\ell(E') & \underset{f'_*}{\overset{f'^{\#}}{\rightleftarrows}} & s\ell(E) \\ {\scriptstyle p'^{\#}}\uparrow\downarrow{\scriptstyle p'_*} & & {\scriptstyle p_*}\downarrow\uparrow{\scriptstyle p^{\#}} \\ s\ell(S') & \underset{f_*}{\overset{f^{\#}}{\rightleftarrows}} & s\ell(S) \end{array}$$

the canonical natural transformation

$$f^{\#}p_* \to p'_*f'^{\#}$$

is an equivalence.

In this situation, we say the Beck-Chevalley condition is satisfied. Notice that, once the result is established, it will also hold with $s\ell$ replaced by Loc, since the functors involved are the same.

Proof: We are assuming that E is the category of sheaves on some site in S, so P can be factored in the form

$$\begin{array}{ccc} sh(B) & \searrow & \\ {\scriptstyle p}\downarrow & & S^{\underline{A}^{op}} \\ S & \swarrow & \end{array}$$

where $\underline{A}$ is a finitely complete category in S, and $B \in \mathrm{Loc}(S^{\underline{A}^{op}})$. We treat each case separately, using the results of §2 and §3.

First, let $\underline{A}$ be a finitely complete category in S, and consider the pullback square

$$\begin{array}{ccc} S'^{f^*\underline{A}^{op}} & \xrightarrow{f'} & S^{\underline{A}^{op}} \\ {\scriptstyle p'}\downarrow & & \downarrow{\scriptstyle p} \\ S' & \xrightarrow{f} & S \end{array}.$$

Let $M \in s\ell(S'^{f^*\underline{A}^{op}})$ and $A \in \underline{A}$. Since

$$(f'_*M)(A) = f_*(M(A)),$$

we must have

$$(f'^{\#}N)(A) \simeq f^{\#}(N(A))$$

for $N \in s\ell(S^{\underline{A}^{op}})$. In fact, $f^{\#}(N(A))$ defines a sup-lattice of $S'^{f^*\underline{A}^{op}}$ by Proposition 1 §2, and the functor from $s\ell(S^{\underline{A}^{op}})$ to $s\ell(S'^{f^*\underline{A}^{op}})$ so defined is clearly a left adjoint of f_*, so the claim follows by uniqueness of adjoints.

Now, however, if $N \in s\ell(S^{\underline{A}^{op}})$, we have

$$f^{\#}p_*(N) = f^{\#}(N(1)) = p'_*f'^{\#}(N),$$

and the result is established in this case.

Next, let $A \in Loc(S)$ and consider the pullback square (Proposition 4 §1)

$$\begin{array}{ccc} sh(f^{\#}A) & \xrightarrow{f'} & sh(A) \\ {\scriptstyle p'}\downarrow & & \downarrow{\scriptstyle p} \\ S' & \xrightarrow{f} & S \end{array} .$$

By Proposition 1 §3,

$$s\ell(sh(A)) \simeq Mod(A).$$

Making the identification, $f'^{\#}$ becomes simply $f^{\#}$ itself, which takes A-modules to $f^{\#}$A-modules. If M is an A-module, p_*M is just M regarded as a sup-lattice, so the Beck-Chevalley condition becomes trivial, and we have proved Theorem 1.

5. The spatial reflection

Let Top/S denote the category of topoi and geometric morphisms over S. Taking sheaves on a space defines a functor

$$sh\colon Sp(S) \to Top/S.$$

We may assign to a topos $p\colon E \to S$ that space whose lattice of open parts is $p_*(\Omega_E)$. The fact, used in the proof of Proposition 4 §1, that if $A \in Loc(S)$ the geometric morphisms

$$\begin{array}{ccc} E & \longrightarrow & sh(A) \\ & {\scriptstyle p}\searrow \quad \swarrow & \\ & S & \end{array}$$

are in one to one correspondence with the locale morphisms

$$A \to p_*(\Omega_E)$$

is nothing but the statement that this assignment defines a reflection of Top/S into Sp(S). We call it the <u>spatial reflection</u>.

As a consequence of the existence of such a reflection, an arbitrary geometric morphism $p\colon E \to S$ can be factored in the form

$$\begin{array}{ccc} E & \xrightarrow{q} & F \\ & {\scriptstyle p}\searrow \quad \swarrow{\scriptstyle r} & \\ & S & \end{array}$$

where r is <u>spatial</u>, i.e. F is sh(X) for $X \in \mathrm{Sp}(S)$, and q is <u>hyperconnected</u>, meaning $q_*(\Omega_E) \simeq \Omega_F$. Clearly, such a factorization is unique, for the space X can only be the space associated to $p_*(\Omega_E)$. We call this the <u>hyperconnected factorization</u> of p. It has also been considered by Johnstone, who, by a totally different method, independently proved the following Proposition in [13].

<u>Proposition 1</u>. The hyperconnected factorization is preserved by pullback.

<u>Proof</u>: We know that the pullback of a spatial morphism is spatial, so it is enough to show that the pullback of a hyperconnected morphism is hyperconnected. So, consider a pullback square

$$\begin{array}{ccc} E' & \xrightarrow{f'} & E \\ {\scriptstyle q'}\downarrow & & \downarrow{\scriptstyle q} \\ S' & \xrightarrow{f} & S \end{array}$$

where q is hyperconnected. By Theorem 1 §4, and Proposition 2 §1, we have:

$$\begin{aligned} q'_*\Omega &= q'_*f'^{\#}\Omega \\ &= f^{\#}q_*\Omega \\ &= f^{\#}\Omega = \Omega . \end{aligned}$$

<u>Proposition 2</u>. The spatial reflection preserves finite products.

<u>Proof</u>: Let $p_1\colon E_1 \to S$ and $p_2\colon E_2 \to S$ be two geometric morphisms. Factor them as

$$\begin{array}{ccc} E_i & \xrightarrow{q_i} & \mathrm{sh}(A_i) \\ & {\scriptstyle p_i}\searrow \quad \swarrow & \\ & S & \end{array}$$

$i = 1,2$, $A_i \in \mathrm{Loc}(S)$, q_i hyperconnected. Now pull back in stages, obtaining the diagram

$$\begin{array}{ccccc}
\mathcal{E}_1\times\mathcal{E}_2 & \xrightarrow{q_1''} & \mathrm{sh}(P_2^{\#}A_1) & \longrightarrow & \mathcal{E}_2 \\
{\scriptstyle q_2''}\downarrow & & {\scriptstyle q_2'}\downarrow & & \downarrow{\scriptstyle q_2} \\
\mathrm{sh}(P_1^{\#}A_2) & \xrightarrow{q_1'} & \mathrm{sh}(A_1\otimes A_2) & \longrightarrow & \mathrm{sh}(A_2) \\
\downarrow & & \downarrow & & \downarrow \\
\mathcal{E}_1 & \xrightarrow{q_1} & \mathrm{sh}(A_1) & \longrightarrow & S
\end{array}$$

which proves the result by Proposition 1 (a composite of hyperconnected morphisms is clearly hyperconnected).

CHAPTER VII - OPEN MORPHISMS OF TOPOI

Here we introduce the notion of an open map of topoi, and study its basic properties. In particular, we establish several equivalent conditions, show openness is stable under pullback, and characterize it in terms of a site in S. Next we prove that any topos can be spatially covered by an open surjection. Finally, we use openness to characterize atomic topoi in analogy with discrete spaces: $p: A \to S$ is atomic iff p and the diagonal $\Delta: A \to A\times A$ are open. Open morphisms have also been considered Mikkelsen [16] (under a different name) and by Johnstone, who independently proved our first three Propositions in [12].

1. Open geometric morphisms

Definition. We say $p: E \to S$ is open iff its spatial reflection is an open space, i.e. iff the canonical morphism of locales

$$P(1) \underset{p^-}{\overset{\exists}{\rightleftarrows}} p_*\Omega$$

has a left adjoint $\exists \dashv p^-$.

Proposition 1. The following conditions are equivalent:

1) $p: E \to S$ is open.

2) For any open space X of E, $p!X$ is an open space of S.

3) If $\begin{array}{c} S \\ \downarrow \\ I \xrightarrow{\alpha} J \end{array}$ is a diagram in S, then

$$p^*(\forall_\alpha S) = \forall_{p^*\alpha} p^*S,$$

i.e. $p^*: S \to E$ preserves universal quantification.

Proof: If p is open and X is an open space in E, then the canonical map $\Omega \to O(X)$ has a left adjoint. p_* is a strong functor, so it preserves the adjointness. Composing with the adjunction $\exists \dashv p^-$ shows that $p!X$ is open. On the other hand, 1 is an open space in E, and $O(p!1) = p_*\Omega$, so $2) \Longrightarrow 1)$. For $1) \Longleftrightarrow 3)$, notice first that we have $1) \Longleftrightarrow$ for each $I \in S$, the canonical map

$$p_I^-: (P1)^I \to p_*(\Omega^{p^*I}),$$

taking a subset $S \rightarrowtail I$ to $p^*S \rightarrowtail p^*I$, has a left adjoint $\exists_I$. This is true because $p_*(\Omega^{p^*I}) \simeq (p_*\Omega)^I$. If $\alpha\colon I \to J$, consider the diagram

$$\begin{array}{ccc} (P1)^I & \overset{\exists_I}{\underset{p_I^-}{\rightleftarrows}} & p_*(\Omega^{p^*I}) \\ (P1)^\alpha \uparrow\downarrow \forall_\alpha & & p_*(\Omega^{p^*\alpha}) \uparrow\downarrow p_*(\forall_{p^*\alpha}) \\ (P1)^I & \overset{\exists_J}{\underset{p_J^-}{\rightleftarrows}} & p_*(\Omega^{p^*J}) \end{array} .$$

From the previous remark, it follows that if $\exists_I$ and $\exists_J$ exist for all I and J, then the part of this diagram involving them, $(P1)^\alpha$, and $p_*(\Omega^{p^*\alpha})$ commutes. But in that case, the rest of the diagram, consisting of their right adjoints, must also commute, which is just 3). If 3) holds, then each p_I^- commutes with infima, since these are an instance of universal quantification. But then we have $\exists_I$ for each I, completing the proof.

Notice that condition 2) shows any $p\colon E \to S$ is open when S is boolean, for any space in a boolean topos is open -- use Proposition 2 §3 Chapter V, together with the well-known fact that any subspace of 1 in a boolean topos is open.

We say $p\colon E \to S$ is an <u>open surjection</u> iff its spatial reflection is - i.e. iff in

$$P(1) \overset{\exists}{\underset{p^-}{\rightleftarrows}} P_*\Omega,$$

$\exists p^- = \mathrm{id}$. We have seen this is true iff p^- is monic, which is the same as saying p is a surjection in the usual sense, i.e. $p^*\colon S \to E$ is faithful.

<u>Proposition 2</u>. In a commutative diagram

$$\begin{array}{ccc} E' & \xrightarrow{f} & E \\ & {\scriptstyle p'}\searrow \quad \swarrow{\scriptstyle p} & \\ & S & \end{array} ,$$

if p and f are open so is p'. Also, if p' is open and f is a surjection, then p is open.

<u>Proof</u>: The first statement is obvious, and the second follows immediately from condition 3) of Proposition 1.

<u>Proposition 3</u>. Open (surjective) morphisms are stable under pullback.

Proof: By Proposition 1 §5 chapter VI, the hyperconnected factorization

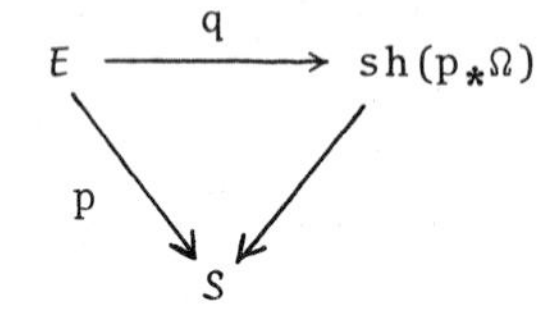

is preserved by pullback. The result now follows from Propositions 3 and 4 of Chapter VI §1.

2. A site characterization of openness

Let $\underline{A}$ be a site in S. More precisely, we consider a covering system on a category $\underline{A}$ in S. Such a system is given by specifying, for each A in $\underline{A}$, a collection $\mathrm{Cov}(A)$ of covers $R = \{A_i \to A\}_{i \in I}$ of A. Let $\hat{R}$ denote the sieve generated by R: $\alpha\colon A' \to A$ is in $\hat{R}$ iff $\alpha = A' \to A_i \to A$ for some $A_i \to A$ in R. The single condition these covers should satisfy is: if $\alpha\colon A' \to A$ and $R \in \mathrm{Cov}(A)$, then $\{\beta\colon A'' \to A' \mid \alpha\beta \in \hat{R}\} \supseteq R'$ for some $R' \in \mathrm{Cov}(A')$. As in the case of a partially ordered set, a covering system is not yet a topology, since we have not required that identities be covers, or that the local axiom is satisfied. As is now well-known, however, the sheaves on the generated topology are the same as the sheaves on the covering system, so we will not generally distinguish between covering systems and sites.

Definition. A site $\underline{A}$ is said to be open iff for each $A \in \underline{A}$ and each $R = \{A_i \to A\}_{i \in I}$ in $\mathrm{Cov}(A)$, $\exists i \in I$.

Proposition 1. $p\colon E \to S$ is open iff E is $sh(\underline{A})$ for some open site $\underline{A}$ in S.

Proof: Let $\underline{A}$ be a site, and let P be the partially ordered reflection of $\underline{A}$. That is, the elements of P are the objects of $\underline{A}$, and $A' \leq A$ in P iff $A' \to A$ in $\underline{A}$. Define covers of elements of P to be images of covers of objects of $\underline{A}$ - i.e. $\{A_i \leq A\}_{i \in I}$ covers A iff there is a cover $\{A_i \to A\}_{i \in I}$ in $\mathrm{Cov}(A)$. This is a covering system on P, and forming the associated locale quotient Q of $\mathcal{D}(P)$, we obtain, by the results of Chapter III §4,

$$Q = \{S \in \mathcal{D}(P) \mid \forall A \in P\ \forall\, R \in \mathrm{Cov}(A),\ R \subseteq S \Longrightarrow A \in S\}$$

But such S are nothing but the subsheaves of 1 in $S^{\underline{A}^{op}}$, so if $p\colon sh(A) \to S$ is the canonical morphism,

$$p_*\Omega = Q.$$

Now if $\underline{A}$ is open, then $p_*\Omega$ represents an open space by Proposition 2 Chapter V §3.

On the other hand, suppose $p: E \to S$ is open, so $p^-: P(1) \to p_*\Omega$ has a left adjoint $\exists$. Let

$$\{A_i | i \in I\}$$

be a family of generators for E over S, and put

$$J = \{i \in I | \exists\sigma(A_i) = 1\},$$

where $\sigma(A_i)$ is the support of A_i. We claim $\{A_i | i \in J\}$ is also a family of generators for E. For, suppose $\{A_i \to X\}_{i \in S \subseteq I}$ covers X in E. We have

$$P(1) \underset{\longrightarrow}{\overset{\exists}{\longleftarrow}} p_*\Omega \underset{\longrightarrow}{\overset{p_*\sigma}{\longleftarrow}} p_*\Omega^X,$$

and $X = \bigvee_{i \in S} \text{im}(A_i)$. By lemma 2 Chapter V §5, if

$$S' = \{i \in S | \exists\sigma(\text{im } A_i) = \exists\sigma(A_i) = 1\}$$

then $X = \bigvee_{i \in S'} \text{im}(A_i)$.

If $\{A_i \to A\}_{i \in S \subseteq J}$ is a cover for this new family of generators, we have

$$A = \bigvee_{i \in S} \text{im}(A_i),$$

and applying $\exists\sigma$ we see

$$1 = \bigvee_{i \in S} 1 = [\![\exists i \in S]\!]$$

so this is an open site of definition for E.

We remark that this site characterization of openness provides another proof of the stability of open maps under pullback - it is clearly preserved by inverse image.

3. The spatial cover

Here we prove that any topos $p: E \to S$ can be covered by an open surjection $X \to E$, with X spatial over S. To do this, we could take for X the Diaconescu cover of E [8], for that is, in fact, an open surjection. However, we believe that the present proof provides more conceptual insight, illustrating, in addition, an important technique in topos theory summarized by the statement of Proposition 1 below.

Consider first an arbitrary locale $A \in \text{Loc}(S)$. We know that

$sh(A) \to S$ classifies <u>points of</u> $\underline{A}$, i.e. locale morphisms $A \to P(1)$. If A is of the form $L(X)$ - the free locale on $X \in S$ - then $sh(L(X)) \to S$ classifies locale morphisms $L(X) \to P(1)$, which are mappings $X \to P(1)$ in S, i.e. subsets of X. Subtopoi of $sh(L(X))$ correspond to topologies in $sh(L(X))$, or local operators in S on $L(X)$. These correspond to quotient locales of $L(X)$, whose points represent <u>conditions</u> on the subsets of X. Thus, since every locale A is a quotient of a free locale, we may say that topoi of the form $sh(A)$ classify exactly <u>bounded structure</u>.

Recall that if S_{fin} is the topos of (cardinal) finite sets of S, then $S^{S_{fin}}$ classifies objects in topoi over S. It is denoted by $S[X]$.

<u>Proposition 1</u>. Any topos $p: E \to S$ is spatial over $S[X]$.

<u>Proof</u>: Represent E as $sh(\underline{A})$ for $\underline{A}$ a site in S. Then E is the classifying topos for covariant, flat functors on $\underline{A}$, which take covers to epimorphic families - the <u>points</u> of $sh(\underline{A})$. What we do is to sketch the construction of this classifying topos, proceeding in such a way that the result is evidently spatial over $S[X]$.

Thus, let $A_1 \overset{\partial^1}{\underset{\partial^0}{\rightrightarrows}} A_0$, etc., be the underlying category of $\underline{A}$, and let $q: S[X] \to S$ be the canonical morphism. Now, introduce a universal mapping $X \to q^*A_0$. This is spatial over $S[X]$ - simply take the free locale $L(X \times q^*A_0)$, and on the universal relation put the condition that it be the graph of a function. The resulting topos is the classifying topos for a family of objects indexed by A_0. Call it $r: S[E] \to S$, where $E \to r^*A_0$ is the universal family. Next, in $S[E]$ we have the diagram

$$\begin{array}{ccc} & & E \\ & & \downarrow \\ r^*A_1 & \overset{r^*\partial^1}{\underset{r^*\partial^0}{\rightrightarrows}} & r^*A_0 \end{array},$$

and we introduce a universal map from the pullback of E along $r^*\partial^0$ to the pullback of E along $r^*\partial^1$. As above, the result is spatial over $S[E]$, hence over $S[X]$. Call it $v: S[F] \to S$, where the universal data in $S[F]$ is

$$\begin{array}{ccccc} (v^*A_1) \times_{v^*\partial^0} F & \xrightarrow{\alpha} & (v^*A_1) \times_{v^*\partial^1} F & & F \\ & \searrow & \swarrow & & \downarrow \\ & & v^*A_1 & \overset{v^*\partial^1}{\underset{v^*\partial^0}{\rightrightarrows}} & v^*A_0 \end{array}$$

The axioms stating that this is a flat functor taking covers to epimorphic families are all <u>conditions</u> on this data, and yield a subtopos of $S[F]$, which is the category of sheaves on $\underline{A}$.

If in Proposition 1 we assume that $\underline{A}$ has a terminal object, as we certainly may, then from the flatness of the universal functor $F \to p^*A_0$, it follows that the category determined by F also has a terminal object. Hence F is <u>non-empty</u>. As a result, the spatial morphism $sh(\underline{A}) \to S[X]$ constructed above actually factors, necessarily spatially, thru the subtopos $S[U] \hookrightarrow S[X]$, where $U \twoheadrightarrow 1$ is the universal non-empty object.

<u>Theorem 1</u>. For any topos $p: E \to S$ there is an open surjection $X \to E$, with $X \to S$ spatial.

<u>Proof</u>: Denote the natural numbers by $\mathbb{N}$, and in $S[U]$ introduce a universal epimorphism $\mathbb{N} \twoheadrightarrow U$. Call the result $S[\mathbb{N} \twoheadrightarrow U] \to S[U]$. It is spatial, but, moreover, by the example following Proposition 2 §3 Chapter V, it is an open surjection. $S[\mathbb{N} \twoheadrightarrow U] \to S$ classifies arbitrary quotients of $\mathbb{N}$, so it is also the classifying topos for an equivalence relation on $\mathbb{N}$, hence spatial. By the remark following Proposition 1, we have a spatial morphism $E \to S[U]$. Consider the diagram

$$\begin{array}{ccc} X & \longrightarrow & S[\mathbb{N} \twoheadrightarrow U] \\ \downarrow & & \downarrow \\ E & \longrightarrow & S[U] \\ & {\scriptstyle p}\searrow \quad \swarrow & \\ & S & \end{array}$$

where the square is a pullback. $X \to E$ is an open surjection by Proposition 3 §1, and $X \to S$ is spatial since both factors of the composite $X \to S[\mathbb{N} \twoheadrightarrow U] \to S$ are.

4. <u>A characterization of atomic topoi</u>

Here we characterize atomic topoi in a manner analogous to, and using, the characterization of discrete spaces in Chapter IV §5.

<u>Proposition 1</u>. Any point

$$x: S \to A$$

of an atomic topos A over S is open. If A is connected, x is surjective.

<u>Proof</u>: Let $p: A \to S$ be atomic, i.e. p^* is logical, and suppose $x: S \to A$ is a point of A. Let $\underline{A}$ be the category of atoms of A. Then $A = sh(\underline{A})$, and x is induced by a flat functor

$$F\colon \underline{A} \to S$$

taking covers to epimorphic families. If $A \in \underline{A}$, then

$$\frac{\dfrac{A \to x_* P(1)}{x^*A \to P(1)}}{FA \to P(1)}$$

so $x_*[P(1)](A) = P(1)^{FA}$, $\Omega(A) = P(1)$, and $p^-\colon \Omega \to x_* P(1)$ at A is the usual

$$p_A^-\colon P(1) \to P(1)^{FA},$$

which has the standard left adjoint $\exists_A$. In general, $\exists_A$ is not functorial, i.e. if $\alpha\colon A' \to A$ then

$$\begin{array}{ccc} P(1)^{FA} & \xrightarrow{P(1)^{F\alpha}} & P(1)^{FA'} \\ & {\scriptstyle \exists_A} \searrow \quad \swarrow {\scriptstyle \exists_{A'}} & \\ & P(1) & \end{array}$$

does not commute. But here, α is a cover, so $F\alpha$ is epimorphic. Thus, in the pullback

$$\begin{array}{ccc} S' & \twoheadrightarrow & S \\ \downarrow & & \downarrow \\ FA' & \xrightarrow[F\alpha]{} \!\!\!\!\!\to & FA \end{array}$$

$\exists_{A'} S' = \sigma(S') = \exists_A S = \sigma(S)$, and $\exists \dashv p^-$ in A.

If A is connected, 1 is an atom, so each $A \to 1$ is a cover, and $FA \twoheadrightarrow F1 = 1$. Hence $\exists p^- = \mathrm{id}$, and x is surjective.

Theorem 1. $p\colon A \to S$ is atomic iff p and the diagonal $\Delta\colon A \to A\times A$ are open.

Proof: If A is atomic over S, then p^* is logical, so p is open. By Proposition 1, any point of A is open. But pullbacks of atomic topoi are atomic, thus

$$p_1\colon A\times A \to A$$

is atomic, and it has a canonical point

$$\Delta\colon A \to A\times A,$$

which must be open.

On the other hand, suppose p and Δ are open. The spatial reflection preserves finite products (Proposition 2 §5 Chapter VI), hence it preserves the diagonal. It is immediate that the spatial reflection of an open map is open, so the spatial reflection of $p: \mathcal{A} \to \mathcal{S}$ is open with open diagonal, hence discrete by Theorem 1 §5 Chapter IV - i.e. as a space, $p_*\Omega$ is discrete. Now we use the following remark.

Lemma 1. Let $\mathcal{F} \xrightarrow{q} \mathcal{E} \xrightarrow{p} \mathcal{S}$ be a composite geometric morphism. If $\mathcal{F}$ is open with open diagonal over $\mathcal{E}$, and $\mathcal{E}$ is open with open diagonal over $\mathcal{S}$, then $\mathcal{F}$ is open with open diagonal over $\mathcal{S}$.

Proof: The openness of the composite is clear, and the diagonal of $\mathcal{F}$ over $\mathcal{S}$ is the composite

$$\mathcal{F} \xrightarrow{\Delta} \mathcal{F} \underset{\mathcal{E}}{\times} \mathcal{F} \longrightarrow \mathcal{F} \underset{\mathcal{S}}{\times} \mathcal{F} .$$

The first morphism is open by assumption, and the second is the pullback

$$\begin{array}{ccc} \mathcal{F} \underset{\mathcal{E}}{\times} \mathcal{F} & \longrightarrow & \mathcal{E} \\ \downarrow & & \downarrow \\ \mathcal{F} \underset{\mathcal{S}}{\times} \mathcal{F} & \xrightarrow{q \times q} & \mathcal{E} \underset{\mathcal{S}}{\times} \mathcal{E} \end{array} ,$$

which is also open.

Going back to the proof of Theorem 1, it follows from Lemma 1 that

$$\mathcal{A}/X \to \mathcal{A} \xrightarrow{p} \mathcal{S}$$

is open with open diagonal for each $X \in \mathcal{A}$. Thus, from the first part of the proof, each $p_*(\Omega^X)$ represents a discrete space, and we are finished by Proposition 5 §1 Chapter VI.

CHAPTER VIII - DESCENT MORPHISMS OF TOPOI

In this final chapter we study the problem of descending sheaves along a geometric morphism of topoi. We define effective descent morphisms, then prove the basic theorem that among these are the open surjections. Our main theorems on the structure of topoi follow.

1. Effective descent morphisms

Let $f\colon \mathcal{E}' \to \mathcal{E}$ be a geometric morphism over $\mathcal{S}$, and consider the diagram

$$\begin{array}{ccccccc} & \xrightarrow{p_{23}} & & & & & \\ & \xrightarrow{p_{13}} & & \xrightarrow{p_2} & & f & \\ \mathcal{E}'\times_{\mathcal{E}}\mathcal{E}'\times_{\mathcal{E}}\mathcal{E}' & \xrightarrow[p_{12}]{} & \mathcal{E}'\times_{\mathcal{E}}\mathcal{E}' & \xrightarrow[p_1]{} & \mathcal{E}' & \longrightarrow & \mathcal{E}. \\ & & \uparrow{\scriptstyle \Delta} & & & & \\ & & \mathcal{E}' & & & & \end{array}$$

Descent data on $X' \in \mathcal{E}'$ is a morphism

$$\theta\colon p_1^*X' \to p_2^*X'$$

such that

$$\Delta^*\theta = \mathrm{id}_{X'}$$

$$p_{13}^*\theta = p_{23}^*\theta \circ p_{12}^*\theta .$$

$\mathrm{Des}(f)$ is the category of objects of $\mathcal{E}'$ equipped with descent data and morphisms which respect the descent data in the obvious sense. Clearly, f^* induces a functor $\Phi\colon \mathcal{E} \to \mathrm{Des}(f)$ such that

$$\begin{array}{ccc} \mathcal{E} & \xrightarrow{\Phi} & \mathrm{Des}(f) \\ & {\scriptstyle f^*}\searrow \quad \swarrow{\scriptstyle U} & \\ & \mathcal{E}' & \end{array}$$

commutes, where $U(X',\theta) = X'$. We say f is a descent morphism of topoi when Φ is full and faithful, and an effective descent morphism when Φ is an equivalence.

Notice that descent data is a morphism $\theta\colon p_1^*X' \to p_2^*X'$, which is equivalent to a morphism $\overline{\theta}: X' \to p_{1*}p_2^* X'$. However, the canonical natural transformation $f^*f_* \to p_{1*}p_2^*$ is <u>not</u> an equivalence in general, i.e. the Beck-Chevalley condition is not satisfied, so we cannot equate descent data with an f^*f_*-coalgebra structure as we did in the algebraic case.

2. Open surjections

Here we prove our principal result:

Theorem 1. Open surjections are effective descent morphisms.

Proof: Let $f: \mathcal{E}' \to \mathcal{E}$ be an open surjection. Consider first the spatial case: $\mathcal{E}' = sh(X)$, $\mathcal{E} = sh(Y)$, so f comes from a continuous, open, surjective map $f: X \to Y$ of spaces in $\mathcal{S}$. We are trying to descend objects of $\mathcal{E}'$, i.e. sheaves on X, i.e. etale spaces $E \xrightarrow{p} X$. We have

$$\mathcal{O}(Y) \underset{f^-}{\overset{\exists_f}{\rightleftarrows}} \mathcal{O}(X)$$

such that $\exists_f \dashv f^-$, $\exists_f$ is $\mathcal{O}(Y)$ linear, and $\exists_f f^- = \text{id}$. f^- is an effective descent morphism for $\mathcal{O}(X)$-modules by Theorem 2 §5 Chapter II. Etale spaces $p: E \to X$ correspond to locale extensions $p^-: \mathcal{O}(X) \to \mathcal{O}(E)$ such that p^- and the binary infimum $\mathcal{O}(E) \underset{\mathcal{O}(X)}{\otimes} \mathcal{O}(E) \to \mathcal{O}(E)$ have linear left adjoints. Now we know that commutative $\mathcal{O}(X)$-algebras descend by the remarks at the end of Chapter II §5. To check that locale extensions descend, we must show that if $\mathcal{O}(Y) \to A$ is a commutative $\mathcal{O}(Y)$-algebra such that $\mathcal{O}(X) \underset{\mathcal{O}(Y)}{\otimes} A$ is a locale, then A was a locale. There is no trouble with the axiom $\forall a \in A,\ a \leq 1$, i.e. $1 = \bigvee_{a\in A} a$, since the two sides of this equation represent two $\mathcal{O}(Y)$-module morphisms $\mathcal{O}(Y) \rightrightarrows A$, which become equal under $\mathcal{O}(X) \underset{\mathcal{O}(Y)}{\otimes} (\ )$, and must therefore have been equal to start with. The axiom $a^2 = a$ is not quite so simple, because the morphism $(\ )^2: A \to A$ exists only in $\mathcal{S}$, not in $\text{Mod}(\mathcal{O}(Y))$. However, it is easy to see that, in $\mathcal{S}$, the diagram

$$\begin{array}{ccc} A & \xrightarrow{\eta A} & \mathcal{O}(X) \underset{\mathcal{O}(Y)}{\otimes} A \\ \downarrow{\scriptstyle (\ )^2} & & \downarrow{\scriptstyle (\ )^2} \\ A & \xrightarrow{\eta A} & \mathcal{O}(X) \underset{\mathcal{O}(Y)}{\otimes} A \end{array}$$

commutes. In fact, $\eta A(a) = 1\otimes a$ and $(1\otimes a)^2 = 1\otimes a^2 = \eta A(a^2)$. ηA is a monomorphism (in fact a retract) in $\mathrm{Mod}(\mathcal{O}(Y))$, hence in S. So, if $(\ )^2 = \mathrm{id}$ on $\mathcal{O}(X) \underset{\mathcal{O}(Y)}{\otimes} A$, $(\ )^2 = \mathrm{id}$ on A. Thus $\mathcal{O}(Y)$ locale extensions descend. Similarly, if a morphism between locale extensions descends, and has a linear left adjoint, then the descended morphism has a linear left adjoint. This is true because the linear left adjoint is automatically compatible with the descent data, and $\mathcal{O}(X) \underset{\mathcal{O}(Y)}{\otimes} (\ )$ preserves and reflects the order relation on morphisms, so the adjoint descends. As a result, sheaves descend uniquely, and the theorem is proved in the spatial case.

Now let $f: E' \to E$ be an arbitrary open surjection. Take the spatial cover $q: X \to E'$, which is an open surjection by Theorem 1 §3 Chapter VII. Consider the diagram

$$\begin{array}{ccc} X \underset{E}{\times} X & \xrightarrow{q\times q} & E' \underset{E}{\times} E' \\ \pi_1 \Big\downarrow\Big\downarrow \pi_2 & & p_1 \Big\downarrow\Big\downarrow p_2 \\ X & \xrightarrow{q} & E' \\ & \searrow^{u} & \Big\downarrow f \\ & & E \end{array}$$

where $u = fq$. Suppose $X' \in E'$ has descent data $\theta: p_1^* X' \to p_2^* X'$ on it. Then in

$$\begin{array}{ccc} (q\times q)^* P_1^* X' & \xrightarrow{(q\times q)^*\theta} & (q\times q)^* P_2^* X' \\ \simeq \Big\downarrow & & \simeq \Big\downarrow \\ \pi_1^*(q^* X') & \longrightarrow & \pi_2^*(q^* X') \end{array} ,$$

$(q\times q)^*\theta$ is descent data on q^*X'. u is an open surjection and spatial over E, hence an effective descent morphism by the above. Thus, there is $X \in E$ and an isomorphism

$$q^*X' \xrightarrow[\alpha]{} u^*X \simeq q^*(f^*X)$$

compatible with the descent data. But X is spatial over E, hence spatial over E', and q is an open surjection, so q is an effective descent morphism, and α descends uniquely to an isomorphism $X' \xrightarrow[\simeq]{} f^*X$, which itself is compatible with the original descent data because $q\times q$ is an (open) surjection. Thus, f is an effective descent morphism, and the proof is finished.

<u>Theorem 2</u>. $p: E \to S$ is an effective descent morphism iff its spatial reflection is.

Proof: Let $\mathcal{X} \xrightarrow{r} \mathcal{S}$ in

$$\begin{array}{ccc} \mathcal{E} & \xrightarrow{q} & \mathcal{X} \\ & {\scriptstyle p}\searrow \quad \swarrow{\scriptstyle r} & \\ & \mathcal{S} & \end{array}$$

be the spatial reflection of $\mathcal{E}$. q is hyperconnected, hence an open surjection, hence an effective descent morphism by Theorem 1. If p is an effective descent morphism, then the same proof as in the general case of Theorem 1 shows that r is.

On the other hand, suppose r is an effective descent morphism. We give a sketch of the proof, leaving details to the reader. Consider the diagram

$$\begin{array}{ccccc} & & \mathcal{E}\times_{\mathcal{S}}\mathcal{E} & \xrightarrow{q\times q} & \mathcal{X}\times_{\mathcal{S}}\mathcal{X} \\ & {\scriptstyle\delta}\nearrow & {\scriptstyle p_1}\downarrow\downarrow{\scriptstyle p_2} & & {\scriptstyle r_1}\downarrow\downarrow{\scriptstyle r_2} \\ \mathcal{E}\times_{\mathcal{X}}\mathcal{E} & \overset{q_2}{\underset{q_1}{\rightrightarrows}} & \mathcal{E} & \xrightarrow{q} & \mathcal{X} \\ & & & {\scriptstyle p}\searrow \quad \swarrow{\scriptstyle r} & \\ & & & \mathcal{S} & \end{array}$$

and suppose $\theta\colon p_1^*X \to p_2^*X$ is descent data on $X \in \mathcal{E}$. Then $\delta^*\theta$ is descent data on X with respect to q, so there is $Y \in \mathcal{X}$ and a compatible isomorphism $X \simeq q^*Y$. Now

$$\begin{array}{ccc} (q\times q)^*r_1^*Y & & (q\times q)^*r_2^*Y \\ \downarrow\simeq & & \downarrow\simeq \\ p_1^*q^*Y & & p_2^*q^*Y \\ \downarrow\simeq & & \downarrow\simeq \\ p_1^*X & \xrightarrow{\theta} & p_2^*X \end{array}$$

and $q\times q$ is hyperconnected, so $(q\times q)^*$ is full and faithful and there exists $\theta'\colon r_1^*Y \to r_2^*Y$ which is descent data with respect to r on Y. Thus there is an $S \in \mathcal{S}$ and a compatible isomorphism $Y \xrightarrow[\simeq]{} r^*S$. The composite

$$X \xrightarrow[\simeq]{} q^*Y \xrightarrow[\simeq]{} q^*r^*S \simeq p^*S$$

is the isomorphism we need to complete the proof.

3. Applications to the structure of topoi

Let G be a *spatial group*, i.e. G is a group in $Sp(S)$. We say G is *open* if it is open as a space. A *G-space* is a space X equipped with a continuous action $G \times X \to X$ satisfying the usual identities. A *morphism of G-spaces* is a continuous map compatible with the given actions. A *discrete G-space* is one whose underlying space is discrete. We will prove:

Theorem 1. Let $p: A \to S$ be a connected atomic topos with a point $x: S \to A$. Then there is an open spatial group G such that A is equivalent to the category of discrete G-spaces.

To describe the structure of a completely arbitrary topos $p: E \to S$, we must consider the more general notions of a *spatial groupoid* $\underline{G}$, and a *discrete* $\underline{G}$-space. The definition of spatial groupoid is clear - it is a groupoid in $Sp(S)$. To define the latter, suppose $\underline{G}$ is given by a diagram

$$G_1 \times_{G_0} G_1 \; \begin{array}{c} \xrightarrow{\partial^2} \\ \xrightarrow{\partial^1} \\ \xrightarrow{\partial^0} \end{array} \; G_1 \; \begin{array}{c} \xrightarrow{\partial^1} \\ \xleftarrow{i} \\ \xrightarrow{\partial^0} \end{array} \; G_0$$

in $Sp(S)$: $\partial^0, \partial^1: G_1 \to G_0$ are the domain and codomain ($\underline{G}$ is said to be *open* if these are open maps), i gives the identities, and the remaining $\partial^0, \partial^1, \partial^2$ are respectively, the first projection, composition, and the second projection. (There is, of course, also an inverse $G_1 \to G_1$, which we leave unspecified.) By a $\underline{G}$-space we mean simply a $Sp(S)$-valued continuous functor on $\underline{G}$. That is, a space $X \to G_0$ over G_0, together with an action

$$\begin{array}{ccc} \partial^0{*}X & \xrightarrow{\theta} & \partial^1{*}X \\ & \searrow \quad \swarrow & \\ & G_1 & \end{array} ,$$

such that $i{*}\theta = id$, and $\partial^0{*}\theta \circ \partial^2{*}\theta = \partial^1{*}\theta$. A *discrete* $\underline{G}$-space is one for which the map $X \to G_0$ is etale. Our main theorem is the following:

Theorem 2. Let $p: E \to S$ be an arbitrary topos over S. Then there is an open spatial groupoid $\underline{G}$ such that E is equivalent to the category of discrete $\underline{G}$-spaces.

Proof: Consider the diagonal

$$\Delta: E \to E \times E.$$

The product in Top/S is a 2-product, not an ordinary product, so Δ is not a subtopos, but rather classifies the notion of isomorphism between models of the structure classified by E. This is bounded over $E\times E$, so Δ is spatial.

Let $q\colon X \to E$ be the spatial cover of E. It is an effective descent morphism by Theorem 1 §2. The 2-kernel pair of q can be calculated as the pullback of the diagonal Δ along $q\times q$:

$$\begin{array}{ccc} X\times_E X & \longrightarrow & E \\ {\scriptstyle \langle p_1,p_2\rangle}\downarrow & & \downarrow{\scriptstyle \Delta} \\ X\times X & \xrightarrow{q\times q} & E\times E \end{array}$$

By the remarks above, the structure of $\langle p_1,p_2\rangle$ is that of a groupoid in Top/S - note that p_1 and p_2 are open surjections, being the projections in the pullback of q with itself. It is also spatial, as is X itself. Thus, $\langle p_1,p_2\rangle$ is a groupoid in the category of spatial topoi over S. The latter is equivalent to Sp(S), so there is an open spatial groupoid $\underline{G} = G_1 \underset{\partial^0}{\overset{\partial^1}{\rightrightarrows}} G_0$ such that

$$X\times_E X \underset{p_1}{\overset{p_2}{\rightrightarrows}} X = \mathrm{sh}(G_1) \underset{\partial^0}{\overset{\partial^1}{\rightrightarrows}} \mathrm{sh}(G_0).$$

E is equivalent to the category of objects of X - sheaves on G_0 - provided with descent data. But these are etale spaces $X \to G_0$ equipped with a continuous map

$$\begin{array}{ccc} \partial^0{*}X & \xrightarrow{\theta} & \partial^1{*}X \\ & \searrow \quad \swarrow & \\ & G_1 & \end{array}$$

such that $i{*}\theta = \mathrm{id}$ and $\partial^0{*}\theta\circ\partial^2{*}\theta = \partial^1{*}\theta$, i.e. the discrete $\underline{G}$-spaces so we are done.

Proof of Theorem 1. Let $p\colon A \to S$ be a connected atomic topos with a point $x\colon S \to A$. (We should remark that this is not always the case Makkai [15].) x is an open surjection by Proposition 1 Chapter VII §4, and hence an effective descent morphism. Consider the diagram

$$\begin{array}{ccc} S\times_A S & \longrightarrow & A \\ \downarrow & & \downarrow{\scriptstyle \Delta} \\ S \simeq S\times_S S & \xrightarrow{x\times x} & A\times_S A \end{array} .$$

Using the same reasoning as in the proof of Theorem 2, $S \times_A S \to S$ is $sh(G) \to S$ for an open spatial <u>group</u> G, (the space of objects of the groupoid is 1 here), and A is equivalent to the category of discrete G-spaces.

We remark that the group G of theorem 2 is not as abstract as it might at first appear. The point $x: S \to A$ represents a model M in S of the structure classified by A, and G is the spatial group of automorphisms of M.

Finally, let us consider the case of an <u>étendu</u> E over S. By definition, there is a non-empty object $X \twoheadrightarrow 1$ in E such that E/X is spatial over S. Replacing the spatial cover of E by the open surjection $E/X \to E$ in the proof of Theorem 2 yields

<u>Theorem 3</u>. Every étendu E over S can be represented as discrete $\underline{G}$-spaces for a spatial groupoid $\underline{G}$ in which the domain and codomain maps may be taken to be etale.

Over the category of classical sets, and under the assumption of enough points, which makes the spaces classical too, this was stated, without proof, in SGA 4 [2] p. 481.

REFERENCES

[1] M. Artin, Commutative rings, Course Notes, M.I.T. (1966).

[2] M. Artin, A. Grothendieck, J.L. Verdier, Théorie des topos et cohomologie étale des schemas, SGA 4, Springer Lecture Notes in Mathematics 269 (1972).

[3] B. Banaschewski, E. Nelson, Tensor products and bimorphisms, Canadian Math. Bull. 19 (4) (1976).

[4] M. Barr, *-Autonomous categories, Springer Lecture Notes in Mathematics 752 (1979).

[5] M. Barr, R. Diaconescu, Atomic toposes, J. Pure and Applied Algebra 17 (1980) 1-24.

[6] M. Barr, R. Paré, Molecular toposes, J. Pure and Applied Algebra 17 (1980) 127-152.

[7] A. Boileau, A. Joyal, La Logique des topos, J. Symbolic Logic 46 (1981) 6-16.

[8] R. Diaconescu, Grothendieck toposes have boolean points - a new proof, Comm. in Algebra 4 (1976) 723-729.

[9] A. Grothendieck, Revêtements étales et groupe fondamental SGA 1, Springer Lecture Notes in Mathematics 224 (1971).

[10] C.H. Dowker, D. Papert, Quotient frames and subspaces, Proc. London Math. Soc. 16 (1966) 275-296.

[11] J.R. Isbell, Atomless parts of spaces, Math. Scand. 31 (1972) 5-32.

[12] P.T. Johnstone, Open maps of toposes, Manuscripta Mathematica 31 (1980) 217-247.

[13] ____________, Factorization theorems for geometric morphisms 1, Cahiers de Topologie et Géometrie Différentielle XXII (1981) 3-15.

[14] ____________, Stone Spaces, Cambridge University Press 1982.

[15] M. Makkai, Full continuous embeddings of toposes, Trans. AMS 269 (1982) 167-196.

[16] C.J. Mikkelsen, Lattice-theoretic and logical aspects of elementary topoi, Aarhus Univ. Var. Publ. Ser. 25 (1976).

[17] S. Papert, An abstract theory of topological spaces, Proc. Camb. Phil. Soc. 60 (1964) 197-202.

[18] H. Wallman, Lattices and topological spaces, Ann. of Math. 39 (1938) 112-126.

General instructions to authors for
PREPARING REPRODUCTION COPY FOR MEMOIRS

For more detailed instructions send for AMS booklet, "A Guide for Authors of Memoirs." Write to Editorial Offices, American Mathematical Society, P. O. Box 6248, Providence, R. I. 02940.

MEMOIRS are printed by photo-offset from camera copy fully prepared by the author. This means that, except for a reduction in size of 20 to 30%, the finished book will look exactly like the copy submitted. Thus the author will want to use a good quality typewriter with a new, medium-inked black ribbon, and submit clean copy on the appropriate model paper.

Model Paper, provided at no cost by the AMS, is paper marked with blue lines that confine the copy to the appropriate size. Author should specify, when ordering, whether typewriter to be used has PICA-size (10 characters to the inch) or ELITE-size type (12 characters to the inch).

Line Spacing – For best appearance, and economy, a typewriter equipped with a half-space ratchet – 12 notches to the inch – should be used. (This may be purchased and attached at small cost.) Three notches make the desired spacing, which is equivalent to 1-1/2 ordinary single spaces. Where copy has a great many subscripts and superscripts, however, double spacing should be used.

Special Characters may be filled in carefully freehand, using dense black ink, or INSTANT ("rub-on") LETTERING may be used. AMS has a sheet of several hundred most-used symbols and letters which may be purchased for $5.

Diagrams may be drawn in black ink either directly on the model sheet, or on a separate sheet and pasted with rubber cement into spaces left for them in the text. Ballpoint pen is *not* acceptable.

Page Headings (Running Heads) should be centered, in CAPITAL LETTERS (preferably), at the top of the page – just above the blue line and touching it.

LEFT-hand, EVEN-numbered pages should be headed with the AUTHOR'S NAME;

RIGHT-hand, ODD-numbered pages should be headed with the TITLE of the paper (in shortened form if necessary).

Exceptions: PAGE 1 and any other page that carries a display title require NO RUNNING HEADS.

Page Numbers should be at the top of the page, on the same line with the running heads.

LEFT-hand, EVEN numbers – flush with left margin;

RIGHT-hand, ODD numbers – flush with right margin.

Exceptions: PAGE 1 and any other page that carries a display title should have page number, centered below the text, on blue line provided.

FRONT MATTER PAGES should be numbered with Roman numerals (lower case), positioned below text in same manner as described above.

MEMOIRS FORMAT

It is suggested that the material be arranged in pages as indicated below.
Note: Starred items (*) are requirements of publication.

Front Matter (first pages in book, preceding main body of text).

Page i – *Title, *Author's name.

Page iii – Table of contents.

Page iv – *Abstract (at least 1 sentence and at most 300 words).

*1980 Mathematics Subject Classifications represent the primary and secondary subjects of the paper. For the classification scheme, see Annual Subject Indexes of MATHEMATICAL REVIEWS beginning in December 1978.

Key words and phrases, if desired. (A list which covers the content of the paper adequately enough to be useful for an information retrieval system.)

Page v, etc. – Preface, introduction, or any other matter not belonging in body of text.

Page 1 – Chapter Title (dropped 1 inch from top line, and centered).

Beginning of Text.

Footnotes: *Received by the editor date.

Support information – grants, credits, etc.

Last Page (at bottom) – Author's affiliation.

ABCDEFGHIJ–AMS–8987654